Sujit Kirpekar

Turbulent Flows in Hard Disk Drives

Sujit Kirpekar

Turbulent Flows in Hard Disk Drives

A Computational Investigation into Flow Induced Vibrations in Hard Disk Drives

VDM Verlag Dr. Müller

Impressum/Imprint (nur für Deutschland/ only for Germany)
Bibliografische Information der Deutschen Nationalbibliothek: Die Deutsche Nationalbibliothek
verzeichnet diese Publikation in der Deutschen Nationalbibliografie; detaillierte bibliografische
Daten sind im Internet über http://dnb.d-nb.de abrufbar.
Alle in diesem Buch genannten Marken und Produktnamen unterliegen warenzeichen-, marken-
oder patentrechtlichem Schutz bzw. sind Warenzeichen oder eingetragene Warenzeichen der
jeweiligen Inhaber. Die Wiedergabe von Marken, Produktnamen, Gebrauchsnamen,
Handelsnamen, Warenbezeichnungen u.s.w. in diesem Werk berechtigt auch ohne besondere
Kennzeichnung nicht zu der Annahme, dass solche Namen im Sinne der Warenzeichen- und
Markenschutzgesetzgebung als frei zu betrachten wären und daher von jedermann benutzt
werden dürften.

Coverbild: www.purestockx.com

Verlag: VDM Verlag Dr. Müller Aktiengesellschaft & Co. KG
Dudweiler Landstr. 125 a, 66123 Saarbrücken, Deutschland
Telefon +49 681 9100-698, Telefax +49 681 9100-988, Email: info@vdm-verlag.de
Zugl.: Berkeley, University of California, Diss., 2006

Herstellung in Deutschland:
Schaltungsdienst Lange o.H.G., Zehrensdorfer Str. 11, D-12277 Berlin
Books on Demand GmbH, Gutenbergring 53, D-22848 Norderstedt
Reha GmbH, Dudweiler Landstr. 99, D- 66123 Saarbrücken
ISBN: 978-3-8364-7494-8

Imprint (only for USA, GB)
Bibliographic information published by the Deutsche Nationalbibliothek: The Deutsche
Nationalbibliothek lists this publication in the Deutsche Nationalbibliografie; detailed
bibliographic data are available in the Internet at http://dnb.d-nb.de.
Any brand names and product names mentioned in this book are subject to trademark, brand or
patent protection and are trademarks or registered trademarks of their respective holders. The use
of brand names, product names, common names, trade names, product descriptions etc. even
without
a particular marking in this works is in no way to be construed to mean that such names may be
regarded as unrestricted in respect of trademark and brand protection legislation and could thus
be used by anyone.

Cover image: www.purestockx.com

Publisher:
VDM Verlag Dr. Müller Aktiengesellschaft & Co. KG
Dudweiler Landstr. 125 a, 66123 Saarbrücken, Germany
Phone +49 681 9100-698, Fax +49 681 9100-988, Email: info@vdm-verlag.de

Copyright © 2008 VDM Verlag Dr. Müller Aktiengesellschaft & Co. KG and licensors
All rights reserved. Saarbrücken 2008

Produced in USA and UK by:
Lightning Source Inc., 1246 Heil Quaker Blvd., La Vergne, TN 37086, USA
Lightning Source UK Ltd., Chapter House, Pitfield, Kiln Farm, Milton Keynes, MK11 3LW, GB
BookSurge, 7290 B. Investment Drive, North Charleston, SC 29418, USA
ISBN: 978-3-8364-7494-8

Preface

This book is essentially an accurate reproduction of the my doctoral thesis from the University of California, Berkeley (2006). It attempts to perform a comprehensive investigation of turbulent flows in operational hard disk drives and the resulting impact on the structures inside the drive. With the rapid proliferation of hard disk drives into non-traditional applications new demands are being placed on the size, speed and reliability of these drives. There has been a strong demand for higher areal density, faster data transfer rates and better reliability. Higher track densities require in the reduction of the available area to position the read-write head, thereby reducing the allowable tolerance for track-misregistration (TMR). On the other hand, higher data transfer rates require higher speeds of disk rotation, which in turn increase the Reynolds number of the air flow and hence the turbulent excitation of the flow.

Numerical simulations of the turbulent flow of air inside model hard disk drives are reported in this book using a commercial CFD software, CFD-ACE. Even with current supercomputer resources direct numerical simulation (DNS) is not feasible. On the other hand, Reynolds Averaged Navier Stokes (RANS) methods would not capture the essential unsteadiness of the flow. For this reason, large eddy simulation (LES) is the most reliable and accurate method for simulating such flows at reasonable cost. Different flow quantities (velocities, pressure, vorticity) are analyzed, global quantities such as drag on the arm and windage are reported, and the coupled flow structure interaction problem is solved. The pressure and shear stress coupling is done in only one direction, from the flow to the structure. It is observed that rapid vorticity shedding occurs from the sharp corners of the arm, and subsequently this vorticity organizes into turbulent eddies. The highly unsteady wake is transported by the rotating disks and is dissipated along the azimuthal span of the drive.

After reporting the basic flow features, attention is devoted to the accuracy and validity of the results. Firstly, the behavior and accuracy of subgrid scale models (SGS), which form the central core of the LES technique, are investigated. Three different SGS models are compared with a direct numerical simulation. It is shown that the algebraic dynamic model is the optimal choice for the SGS model. Next, three commercial CFD codes CFD-ACE, Fluent and CFX are benchmarked in their ability to solve a standard test problem for LES – the flow across a square cylinder.

To build more credibility into the results, extensive experimental validation is carried out. Validation is carried out against both hot-wire anemometry data (Gross 2003) and particle image velocimetry (Barbier 2006) data. In the context of experimental validation

i

comprehensive grid convergence studies are also performed. It is shown that grids in the 2-2.5 Million cell range are in asymptotic range for convergence. The rates of convergence agree well with the theoretical rates for the discretization schemes. The grid based uncertainty of our results is then estimated to be approximately between 15-30%.

The LES results are then applied to two problems: computation of the flow induced vibrations of the rotating disk and computation of the vibrations of the arm in the presence of flow mitigation devices. In the former, a self developed spectral finite-difference code is used to solve for the elastic vibrations of the rotating disk. In the latter, HDD casings which claim reduction in flow induced vibrations by the use of small geometrical modifications are investigated. These modification include, an upstream spoiler, a downstream spoiler and a blocking plate. It is observed that the blocking plate is the most effective in reducing the flow induced vibrations of the arm.

Finally, a novel and computationally inexpensive technique is suggested as a method for solving flow induced vibration problems. By approximating the flow induced forcing spectrum by a piecewise linear model, it is shown that the computational cost may be reduced significantly without sacrificing much accuracy of the results.

My doctoral work (and hence this book) would not have been possible without the support and encouragement of my parents. I'd like to thank them for their undying love and affection towards me. I also greatly acknowledge my wife and her family, our baby daughter and my brother and his family for the joy and support they give me.

I'd like to thank my research advisor Prof. David B. Bogy not only for the excellent technical advice, but also for the brilliant professional development that I received under him. My sincere thanks to: Vineet Gupta, Nihar Shah, Rohit Karnik, Rohit Ambekar, Amol Phadke, Shrikant Bharadwaj, Brendan Cox, Sean Moseley Puneet Bhargava, Du Chen, Prof. Philip Marcus, Prof. Stanley Berger, Prof. Omer Savas, Prof. Oliver O'Reilly, Prof. Fai Ma, Prof. Andrew J. Szeri, Prof. John Strain, Prof. Grigory Barenblatt (from UC Berkeley), Charlotte Barbier, Prof. J.A.C. Humphrey, Mohammed Kazemi, (from the University of Virginia), Shane Moeykens, Pamela McClay, Chris Lariviere (from Fluent Inc,), John Kiefer, Hoang Vinh(from CFX), Jim Lindauer and the support staff (from ESI Software), Ferdi Hendricks, Andre Chan (from Hitachi GST), Srinivas Tadepalli, Hany Gross (from Seagate), Bob Evans, Mohammed Kazemi (from Hutchinson Technology), Haesung Kwon (from Samsung) and Jen-Tai Lin (from Western Digital).

Sujit Kirpekar
March 2008

To Mum and Dad

iv

Contents

Chapter 1

Introduction

1.1 Introduction

This book is a computational investigation into the turbulent flows that occur inside hard disk drives. In this chapter, we begin by giving the reader a short historical account of the development of disk drives. Several terms that are specific to disk drives are defined, the problem under investigation is outlined and some of the common strategies used to solve the problem are listed. Here, and in the rest of the chapters, *hard disk drive*, *disk drive*, *HDD*, or simply *drive* mean the same and are used interchangeably.

1.1.1 A brief introduction to hard disk drives

The IBM 350 (part of the IBM RAMAC 305) is often claimed to be the first commercial disk drive product, introduced on September 4, 1956. (Wikipedia 2006) RAMAC stood for "Random Access Method of Accounting and Control". The IBM 350 stored 5 million characters (about 5 megabytes), had fifty disks of 24-inch diameter with 100 recording surfaces. The disks spun at 1200 rotations per minute (RPM). Data transfer rate was 8,800 characters per second. Two independent access arms moved up and down to select a disk and in and out to select a recording track, all under servo control. Interestingly, the IBM RAMAC 305 system with the IBM 350 disk storage leased for $3,200 per month.

Exactly fifty years later, at the time of writing of this book, the Seagate ST3500641A disk drive, which has two disks spinning at 7200 RPM, offers a storage capacity of 500 gigabytes. The data transfer rate is about 300 megabytes a second. A quick search on a popular retail website indicates that one can buy this drive for $217! Moreover, the Seagate drive is about $10 \times 15 \times 2.5$ cm in size, which is grossly smaller than the original IBM machines. Figures 1.1 and 1.2 show both these drives.

This simple example really demonstrates the astronomical progress that disk drives have made in the past half century. Moore's Law which is famous in the semiconductor industry, states that at the current rate of technological development, the complexity of an integrated circuit, with respect to minimum component cost, will double in about 18 months. On similar lines, the newly coined "Kryder's Law" [1] states that at current technological advances, the storage density on disk drives will double approximately every 13 months.

Magnetic disk-based storage has been a critical component of the computer revolution and is playing an even more important role today. The exponential increase in storage density (at the same cost) has enabled the commercial viability of consumer products that require large storage capacities, such as the Apple iPod digital music player, the TiVo personal video recorder, and Google's Gmail web-based email program.

1.1.2   Definitions and nomenclature

Disk drives consist of one or more rotating disks on which data is stored on several closely-packed circumferential "tracks". Each track again consists of several thousand "magnetic bits", where data can be stored in the form of a binary unit. Data is read from and written to these bits using a "read/write head", which is usually a giant-magnetoresistive coil that changes the polarity of the magnetic bits. Since the bits are closely packed, the read-write head needs to be positioned very close to the bits. This is done using a "slider" which is either an active or passive element that "flies" a few nanometers over the rotating disk by means of an air bearing. The "actuator" is an internal arm in a disk drive that moves the read/write element from one location to another. Generally, modern actuators are driven by a voice-coil motor and servo control is utilized to accurately position the arm.

Storage density is usually characterized by one of the following metrics: *tracks-per-inch (TPI)*, is the number of tracks that fit into one radial inch; *bits-per-inch (BPI)* is the number of bits in a single track length of one inch; and *areal density (gigabits-per-square-inch)* is simply the number of magnetic bits (usually in gigabits) in an area of one square inch. Figure 1.3 shows the remarkable growth in areal density over the past half century, while Figure 1.4 shows the drop in the unit price of storage over the past few years. Increasing areal density requires the head to be positioned closer to the magnetic bits and this trend is shown in Figure 1.5. The figure clearly shows the decreased spacing between the head and the magnetic media for increasing areal densities. It is foreseeable that such high growth in areal density (20-50% annually) and the decrease in head-disk spacing to such small extremes (3-6 nanometers) will eventually slow down. Nonetheless, promising technologies such as perpendicular recording, thermal or piezo-based slider control, heat-assisted magnetic recording, etc are expected to keep growing the areal density. This is briefly shown in the magnetic media roadmap in Figure 1.6.

---

[1]Kryder's Law is named after Mark Kryder, an engineer currently at Seagate Technology

1.2 Motivation

Given that the disks inside a disk drive are spinning (typically between 5400 to 15000 RPM), the air in the drive gets spun up. This air flow impinges on various components in the drive and causes them to vibrate. Generally, the actuator is the most affected by the flow, since it forms a blunt body obstruction to the circumferential flow. Such flow induced vibrations of the actuator cause it to be displaced from its intended position, a condition that is called *track mis-registration (TMR)*.

Incorrect positioning of the arm leads to errors in the read/write operation, which need to be corrected – hence slowing down the speed of operation of the drive. Moreover, since the flow in largely turbulent, the flow-induced vibrations are generally random and cannot be compensated by the control system in advance.

In modern disk drives, there is a continuing trend for higher speeds of rotation – even above 15,000 RPM. This has mainly been driven by the need for faster data transfer rates between the magnetic bits and the read/write head. Increasing the RPM simply increases the Reynolds number (to be defined for this case later) of the flow causing higher turbulent fluctuations. On the other hand, higher areal densities require very accurate positioning of the head over the track. With the current areal density of approximately 200 gigabits/in^2, the requirement for accuracy in the positioning of the arm (called the TMR budget) is approximately 100 nanometers. It is widely projected that conventional technology will ultimately achieve 1 terabit/in^2 and it is foreseeable that at such areal densities, a track density of 0.5 Million TPI will be required, with each recorded bit being roughly 13 x 50 nanometers (Wood et al. 2002). Under such conditions, the tracking accuracy required is approximately 1.5 nm RMS (root mean square).

These trends clearly manifest the need to minimize the effects of air flow on drive level components inside a disk drive. For this reason the flow field inside an HDD has received attention from the research community over the past few years – and is also the topic of this book.

Interestingly, the flow of air is not the only cause of TMR. Several other sources have been identified as:

- A significant source of positioning error is from the servo controller itself

- Hysteresis at the bearing which holds the rotary actuator is a source of positioning error

- Run-out of the disk, clamping distortions at the inner edge of the disk and manufacturing defects in the spindle are also sources of TMR

- The air bearing over which the slider flies displays highly non-linearly behavior at low spacings and when the slider contacts the disk. The elastic nature of the actuator (weakly) couples the flying dynamics of the head with its horizontal motions that cause TMR. Thus short range forces (e.g. intermolecular) which occur at small fly heights and impact and contact forces when the slider contacts the disk could possibly cause TMR

— Shocks or external vibrations affect the disk and the arm – and cause the arm to lose its positioning accuracy

1.2.1   Some proposed solutions

Several methods / techniques have been proposed to mitigate the TMR problem:

— By reducing the disk diameter and increasing its thickness, which increases the rigidity of the disk, reducing TMR caused by disk flutter and spindle run-out

— By increasing the stiffness of the actuator arm, especially the suspension (which is defined later). This causes the modes of vibration to move to higher frequency ranges, thereby reducing their relative amplitudes

— By achieving better control using a dual stage actuator

— By isolating the drive from external vibration, using fluid spindle bearings and possibly replacing the air with a lower density, non-corrosive gas like Helium (Wood et al. 2002)

— By modifying the air flow in the drive using geometrical features, such that the resultant vibrations are reduced. A few such modifications are studied in Chapter 8.

1.3   Objective

Succinctly, *this book aims to provide accurate and reliable computational solutions for flows occurring in disk drives and whenever feasible, to compute the response of the structures to the flow.*

1.3.1   Organization of the Book

This book is organized as follows:

1. Chapter 1 is completed with a brief review of the prior work in this field

2. Chapter 2 outlines our simulation methodology including: the geometric details of computational model, the grid used to discretize the domain, the initial and boundary conditions used, the numerical methods of the code, and solution strategies that are employed for computing the flow-structure interaction. The chapter describes in detail the fluid mechanics of flows in disk drives and concludes with results showing the response of the arm to the flow. In the remaining chapters the focus is shifted from the fluid mechanics of the flow to different aspects of the simulations.

3. Since the sub-grid scale (SGS) model (defined later) is the key element of a large eddy simulation, Chapter 3 compares and contrasts three different SGS models implemented in the same commercial code. Comparisons are made with a direct numerical simulation.

4. Chapter 4 compares three different commercial codes (and their implementation of four SGS models). This is necessary, since this book is heavily based on results from a commercial code (as are many similar comtemporary CFD-related works) . Instead of using the complex disk drive flow, a simpler test case, flow past a square cylinder, is used for benchmarking.

5. Chapter 5 is an attempt at validating our simulation results with some published experimental hot-wire results of Gross (2003). In this chapter, extensive grid refinement studies are performed, and uncertainties due to the grid and due to numerical dissipation are quantified, in the context of experimental validation.

6. Chapter 6 continues the validation efforts of Chapter 5. Simulations are validation with particle-image-velocimetry based experimental data of Barbier (2006). The role of numerical dissipation is also discussed.

7. Chapter 7 demonstrates the effect of the flow on the rotating disk itself. A self developed finite difference code is used to compute the elastic response of the disk to the flow

8. In Chapter 8, three commonly used "devices" which mitigate the effects of the flow on the arm are tested, computationally. Also, the effect of disks spinning in the opposite direction to the convention is tested.

9. Chapter 9 concludes this work, by summarizing the main findings and listing the significant contributions. A general discussion of future challenges follows, and some ideas are presented to advance the-state-of-the-art. Some simple numerical experiments are also performed to demonstrate the viability of the proposed ideas.

1.4 Literature review

There has been significant experimental, theoretical and numerical research on air flow in hard disk drives over the past 30 years. Several recent Ph.D. theses have reviewed the major accomplishments in this field (chronologically, see Gross (2003), Kazemi (2004) and Barbier (2006)). Nevertheless, a literature review is also presented here briefly, with several updates, especially in the computational results. Results presented in Chapter 2 are largely built on the past work outlined below. In later chapters (3-8) which do not deal exclusively with the flow inside a model HDD, shorter literature reviews relevant to the topic under discussion are presented where necessary.

1.4.1   Experimental research

The experimental work of Lennemann (1974) was one of the first experimental investigations directly focused on disk drives. The author used model disks of diameter between 355.6 - 457 mm running at 710-3600 RPM and used water and aluminum powder for flow visualization. Experiments were performed with and without a slider arm. The author shows the existence of a central laminar core that is rotating slightly slower than the disk and a highly turbulent outer region. The paper also contains an extensive list of prior work related to rotating disks, but not specifically disk drives.

Kaneko et al. (1977) performed similar flow visualization experiments to study the flow between disks with and without a cylindrical shroud. They observed a "bumpy laminar core" that extended from the hub to the mid-radius of the disks, followed by a "more turbulent outer region"

Abrahamson et al. (1989) performed experiments using an acid-base indicator, Bromothymol Blue, in water. Disk speeds were varied from 5-50 RPM, the disk diameter was fixed at 112 cm. They observed three distinct regions of flow: "a solid body inner region near the hub, an outer region dominated by counter rotating vortices and a boundary layer region near the shroud". They reported that decreasing the Ekman number ($Ek = \nu/R^2\Omega$) or increasing the axial spacing between the disks resulted in lesser vortical structures in the outer region and consequently greater overall mixing.

Girard et al. (1995) investigated the effect of an actuator-like rotary arm on the flow field in the drive, using water based flow visualization. Their main conclusions were related to the effect of the arm and the wake it creates.

Tzeng and Humphrey (1991), Schuler et al. (1990) and Usry et al. (1993) performed several laser-Doppler velocimetry experiments of rotating disks with and without an obstruction. They primarily reported mean and RMS values of circumferential velocities and the corresponding frequency content. Usry et al. (1993) also conclude that once the flow separates by flowing over the obstruction, "the flow does not recover within one revolution from the effects of the obstruction".

Experimental research using realistic disk drive configurations for suspensions and sliders has been limited. Yamaguchi et al. (1990) performed hot wire anemometer experiments using a suspension in a uniform and rotating flow. They found no noticeable peaks in the frequency content of the flow and concluded that the flow acts as an aperiodic irregular excitation.

In the Ph.D. thesis of Gross (2003), experimental data in the near vicinity of the e-block arm was made available. Gross also investigated the effect of the thickness of the e-block arm and the shape of it's trailing edge on the airflow and consequently on the flow induced vibrations in the slider. The experimental data set resulting from the work of Gross (2003) has been very useful as a validation tool for our simulations. This validation is presented in Chapter 5.

The thesis of Barbier (2006) is a recent addition to the experimental works. The thesis details results using hot-wire anemometry and particle-image velocimetry. Measurements were made in several locations upstream and downstream of the arm using a twice-large (2x) model of a disk drive. Asymptotic behavior of the velocity profile was observed for increasing speeds of rotation. This dataset is also used for validation in Chapter 6.

1.4.2 Numerical research

Among the first numerical investigations of the air flow in disk drive like enclosures was done by Chang et al. (1990). Using a finite difference code incorporating the $k - \epsilon$ model, they showed good agreement between experiments and simulation with regard to the mean flow velocity and heat transfer characteristics.

The first three-dimensional numerical study of the *unsteady* flow was published by Humphrey et al. (1995). They showed that the toroidal vortices at the shroud "acquire a time-varying sinuous shape in the circumferential direction".

Using a different code, Suzuki and Humphrey (1997) numerically studied the effect of a radially inserted actuator arm and an "airlock" (which is a similar obstruction to the flow). They mainly discuss the pressure, shear stress and disk torque coefficient that they compute. Using the same code as Suzuki and Humphrey (1997), Iglesias and Humphrey (1998) performed 2- and 3-dimensional calculations for different Reynolds numbers. Using a similar non-commercial software Kazemi (2004) has conducted 2-D and 3-D numerical calculations of the flow around a suspension-head unit and reports the resulting vibrations calculated by a finite element technique.

Most of the recent works on air flows in disk drives have used commercial computational fluid dynamics (CFD) software. Due to the rapid increase in computer speeds and research advances in turbulence modeling, numerical investigations are increasingly modeling the geometrical complexities of a real HDD.

Ng and Liu (2001) performed CFD calculations using CFX-5, Shimizu et al. (2001) used large eddy simulation (LES) to study flow induced disk flutter, and Shimizu et al. used LES to study the airflow induced vibrations of the HGA. Tsuda et al. (2003) report DNS results, while Tatewaki et al. (2001) report LES results of airflows in realistic disk drives.

Recognizing that the air flow in a disk drive is highly unsteady and random, most researchers have performed unsteady (time-marching) calculations, typically using LES, (or where resources permit, DNS). Calculations based on Reynolds Averaged methods (which are useful in predicting mean flow fields and particle trajectories) have also been reported by Song et al. (Jan 2004)

Finally, there has also been some published work on reducing flow induced vibrations in disk drives. Hirono et al. (2004) study the effect of an upstream spoiler, while Nakamura et al. (2004) study the effect of miniaturizing the suspension. There has also been some (experimental) work using very similar modifications that have been studied here. E.g. Deeyienyang and Ono (2001) studied the use of "squeeze air bearing plates" in reducing the vibrations of the disk. Other methods of mitigating flow induced vibrations have also been proposed. E.g. Hendriks and Chan (2005) propose the use of an "aerodynamic bypass" which can offer drastic reduction in the upstream pressure at lower costs.

1.5 Figures

Figure 1.1: IBM 305 at U. S. Army Red River Arsenal Foreground: Two 350 disk drives. from Wikipedia (2006)

Figure 1.2: The Seagate Barracuda 7200.9, ST3500641A.

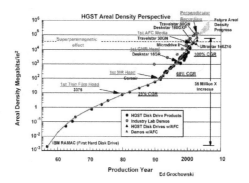

Figure 1.3: Areal Density growth in time; from the first hard drive (1956) to 2004. Courtesy of Hitachi Global Storage Technologies

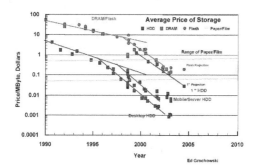

Figure 1.4: Price of storage per unit megabyte, over the past 15 years. Courtesy of Hitachi Global Storage Technologies

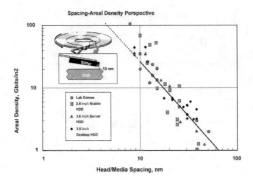

Figure 1.5: Reduction in head-media spacing due to increase areal density. Courtesy of Hitachi Global Storage Technologies

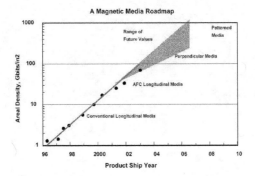

Figure 1.6: Magnetic media roadmap, Courtesy of Hitachi Global Storage Technologies.

Chapter 2

Computing HDD Flows

2.1 Introduction

This Chapter provides a detailed description of the simulation procedures and practices, justifying their use along the way. In addition to the background information about the simulation methodology, this Chapter also serves to describe the fluid mechanics that occur inside realistic disk drive configurations. This Chapter shall serve as a useful guide to the reader wanting to recreate the results described in this and future Chapters.

2.2 CFD Modeling

2.2.1 The finite volume algorithm

Our simulations have been performed using commercial computational fluid dynamics (CFD) software – CFD-ACE. (In Chapter 4, two of the other most popular commercial CFD codes, Fluent and CFX, are tested). The CFD-ACE code includes an unstructured, polyhedral cell flow solver, an interactive geometry modeling and grid generation system and a post-processing system. The code solves the incompressible Navier Stokes equations,

$$\frac{\partial \mathbf{u}}{\partial t} + \mathbf{u} \cdot \nabla \mathbf{u} = -\frac{1}{\rho} \nabla p + \nu \nabla^2 \mathbf{u}; \qquad \nabla \cdot \mathbf{u} = 0 \qquad (2.1)$$

in strong conservation form using the finite volume method. The algorithm used is the well-known SIMPLEC method (semi-implicit method for pressure-linked equations – consistent). A very good description of pressure correction methods is given in Anderson (1995) and the original algorithm is published in Van doormaal and Raithby (1984). We describe the highlights here,

In solving *the integral form* of the Navier Stokes equations (momentum) in discretized form, we seek solutions of the equations of the form,

$$A_P u_{i,P}^{n+1} + \sum_l A_l u_{i,l}^{n+1} = -\left(\frac{\delta p}{\delta x_i}\right)_P^{n+1} \tag{2.2}$$

The above equation represents an equation for the velocity component u_i at point P, $u_{i,l}$ are velocity components at neighboring grid locations, and the coefficients A_P and A_l are determined by the scheme used to discretize the advective and diffusive terms in the Navier Stokes equations. We use the symbol δ to denote the specific numerical scheme to implement the gradient of pressure. Notice that the equations are implicit and hence require the solution of a large system of non-linear equations. We also note, that by choosing an implicit scheme, we are no longer restricted by the CFL-like conditions on the time-step, and our numerical method is assured of unconditional stability in time.

The equation for continuity is represented by,

$$\frac{\delta u_i}{\delta x_i} = 0 \tag{2.3}$$

The SIMPLEC algorithm is inherently iterative – it uses pressure and velocity data from the previous time-step (or iteration) and seeks to correct it by satisfying the continuity and momentum equations. We call the previous values of velocity and pressure by u^{m*} and p^{m*}, and propose corrections of the form,

$$u^m = u^{m*} + u'; \qquad p^m = p^{m*} + p'; \tag{2.4}$$

where the subscript i has been dropped for notational convenience. Since u^{m*} from the previous iteration satisfies Eqn. 2.2 we may write,

$$u_{i,P}^{m*} = \frac{-\sum_l A_l u_l^{m*}}{A_P} - \frac{1}{A_P}\left(\frac{\delta p}{\delta x_i}\right)^{m*} \tag{2.5}$$

or

$$u_{i,P}^{m*} = \tilde{u}_i^{m*} - \frac{1}{A_P}\left(\frac{\delta p}{\delta x_i}\right)^{m*} \tag{2.6}$$

where,

$$\tilde{u}_i^{m*} = \frac{-\sum_l A_l u_l^{m*}}{A_P} \tag{2.7}$$

Taking the divergence of Eqn. 2.6 we obtain the following Poisson equation for the pressure correction,

$$\frac{\delta}{\delta x_i}\left(\frac{\rho}{A_P}\frac{\delta p'}{\delta x_i}\right) = \frac{\delta}{\delta x_i}\left(\rho \tilde{u}_{i,P}'\right) + \frac{\delta}{\delta x_i}\left(\rho u_{i,P}^{m*}\right) \tag{2.8}$$

Since, the term $\tilde{u}_{i,P}'$ is still unknown we approximate it by,

$$\tilde{u}_{i,P}' = -u_{i,P}'\frac{\sum_l A_l}{A_P} \tag{2.9}$$

This gives us the final equation to solve, so that the pressure correction satisfies the divergence condition,

$$\frac{\delta}{\delta x_i} \left(\frac{\rho}{A_P + \sum_l A_l} \frac{\delta p'}{\delta x_i} \right) = \frac{\delta}{\delta x_i} \left(\rho u_{i,P}^{m*} \right) \qquad (2.10)$$

Once the value of pressure correction is obtained, it is used to solve the momentum equations (Eqn. 2.2) to obtain the corrected velocities. In the CFD-ACE code a multi-grid or a conjugate-gradient based method may be used to solve the Possion equation for pressure. This procedure is continued until the corrections obtained are sufficiently small with each iteration. In our simulations the criterion for convergence was maintained at 10^{-4}. Typically, convergence was observed in less than 50 iterations per time step.

2.3 Discretization

The diffusive terms of the Navier Stokes equations are always discretized by second-order central differences. The convective terms pose a harder problem. For the SIMPLEC method described above, fluxes at the cell boundaries need to be evaluated from the variable values at the cell center, for integration of the convective terms. Several methods have been proposed to do this, but we prefer to again use second order central differencing, with the intention of avoiding the well known dissipative errors of upwind-based methods (Mittal and Moin (1997)). However to increase the stability of such a scheme in an inherently iterative solver, sometimes (as in Chapter 4) it is necessary to "blend" the central difference with a first order upwind differencing scheme. The contribution of the upwinding scheme is typically limited to only 10%.

Our time step is chosen so that numerical stability is assured and the turbulent motions are accurately resolved in time. We ensure that our time-step is smaller that the time scale of the smallest resolved scale of motion. This is given by,

$$\tau = \frac{\Delta x}{U} \qquad (2.11)$$

where Δx is an average estimate of the grid size and U is the mean (outer) velocity and that position. Considering this CFL like condition and experience from past research, we choose a time-step of 1×10^{-5} seconds. We also note that the frequencies of oscillations of the structures in a disk drive have experimentally been shown to be of the order of a few kHz, hence such a small time step is indeed necessary to resolve the dynamics of the flow. A time step of 1×10^{-5} seconds allows us to resolve a frequency range up to 50 kHz, which is well above the range of the essential physics.

For advancement in time we use the Implicit Euler's method. Implicit methods are usually needed due to the diffusive terms in the equations of motion. We note that using a first order method will not introduce large errors in the simulation because of the small time step we are using, and the local truncation error is $O(\Delta t^2) = O(10^{-10})$.

2.4   Turbulence Modeling

For disk rotation speeds of 10,000 RPM the linear velocity U of the outer radius of a 3.5 inch disk is 46.54 m/s. Such high speeds generate high shear stresses at the disks, causing large amounts of viscous dissipation. If the disk-to-disk spacing h is 3 mm, the Reynolds number based on the disk spacing, at standard atmospheric conditions (density $\rho = 1.1614$ kg/m3, dynamic viscosity $\mu = 1.864\text{e-}05$ kg/m s), calculated by,

$$Re = \frac{\rho(\Omega R_o)h}{\mu} \tag{2.12}$$

is approximately 8700. Here Ω is the disk angular velocity and R_o is the outer radius of the disk. This is often referred to as the *tip based Reynolds number*. Some authors prefer to report the Reynolds number based on the disk radius, $Re = \frac{\rho(\Omega R_o)R_o}{\mu}$. However, this method leads to higher Reynolds numbers by one order, but is generally not valid as it ignores the length scale in the axial direction. In later Chapters, we only report the tip based Reynolds number.

The Reynolds number of such flows is usually not an accurate indicator of the turbulent nature of the flow. Although the number is seemingly small, the azimuthal symmetry of the flow is broken by the presence of a large obstruction formed by the actuator arm. This blunt body obstruction is a source of turbulent vorticity generation, and any serious modeling effort (as past research has shown) must make use of a turbulence model.

2.4.1   The Kolmogorov microscale

For turbulence modeling we realize that it is not practical to compute the Navier Stokes equations directly, given the complexity of the problem. The Kolmogorov microscale may be computed approximately as,

$$\eta = \left(\frac{\nu^3}{\epsilon}\right)^{1/4} = O(10^{-5})m \tag{2.13}$$

where η is the Kolmogorov's scale, ν is the kinematic (molecular) viscosity and ϵ is the dissipation. Values of ϵ may be easily estimated by a $k-\epsilon$ type Reynolds-Averaged Navier-Stokes (RANS) solution and this calculation implies that our simulation would need approximately $O(10^{10})$ cells in a typical 3D domain. The same result may have been obtained (approximately) by realizing that the number of grid cells in a direct simulation scale as $Re^{9/4}$. From this calculation, we realize that a direct simulation would be impossible given our current computing resources.

While RANS models are not suitable for highly unsteady flows our current workstations necessitate the use of large eddy simulation (LES). Past work also shows that LES has by far been the most practical method to solve flows in disk drives with a reasonable balance between cost and accuracy.

2.5 Large Eddy Simulation (LES)

Turbulent flows consist of a wide range of length and time scales. The larger scales are more energetic than the smaller scales, and they are responsible for the transport of conserved quantities. The smaller scales are universal, self similar and are unaware of the mean flow because such information is lost through the energy cascade procedure (Pope (2003)). Hence large eddy simulation uses a filtering approach to resolve only the larger scales of motion and uses a sub-grid scale (SGS) model to model the unresolved scales. For an incompressible flow, the filtered Navier Stokes equations are (obtained by filtering Eqn. 2.1),

$$\frac{\partial(\rho\overline{u_i})}{\partial x_i} = 0 \tag{2.14}$$

$$\frac{\partial(\rho\overline{u_i})}{\partial t} + \frac{\partial(\rho\overline{u_i u_j})}{\partial x_j} = -\frac{\partial p}{\partial x_i} + \frac{\partial}{\partial x_j}\left[\mu\left(\frac{\partial(\rho\overline{u_i})}{\partial x_j} + \frac{\partial(\rho\overline{u_j})}{\partial x_i}\right)\right] \tag{2.15}$$

where the over bar indicates the filtering operation [1]. The quantity $\overline{u_i u_j} \neq \overline{u_i}\,\overline{u_j}$ on the left side of Eqn. 2.15 is unknown and is replaced by $\overline{u_i}\,\overline{u_j}$. The difference between the terms is modeled by an approximation.

$$\tau_{ij}^R = (\overline{u_i u_j} - \overline{u_i}\,\overline{u_j}) \tag{2.16}$$

Here τ_{ij}^R is called the *sub-grid scale (SGS) stress*, and it represents the interaction of the filtered field with the unresolved field. Different SGS models seek to provide an approximation to the SGS stress term, either through an algebraic equation or by the solution of a differential equation.

The LES research community has produced several SGS models over the past 30 years. Only a few of the available models are used in our work, and hence are described below. For more comprehensive reviews of different SGS models and also for a general introduction to the practice of LES we refer the reader to Ferziger (1983), Ferziger (1996) and Rogallo and Moin (1984).

2.5.1 The Smagorinsky model

The Smagorinsky model (Smagorinsky (1963)) is an algebraic SGS model based on the eddy viscosity hypothesis (gradient diffusion hypothesis) of Boussinesq (1877). Since small scales tend to be more isotropic than large ones it is usually acceptable to parameterize them using an eddy viscosity assumption. The SGS stress is related to the filtered strain rate through a single constant called the eddy viscosity, just as the shear stress is related to the strain rate linearly in a Newtonian fluid. If the filtered strain rate is defined as,

$$\overline{S_{ij}} = \frac{1}{2}\left(\frac{\partial\overline{u_i}}{\partial x_j} + \frac{\partial\overline{u_j}}{\partial x_i}\right) \tag{2.17}$$

---

[1] Filtering involves convolving a quantity with a "filtering kernel" to produce the filtered variable

and the mean strain rate as,

$$|\overline{S}| = \sqrt{2\overline{S_{ij}}\,\overline{S_{ij}}} \qquad\qquad (2.18)$$

the SGS stress (most often only the anisotropic part of the SGS stress) is given by,

$$\tau_{ij}^{R} = -2\nu_T \overline{S_{ij}} \qquad\qquad (2.19)$$

where the eddy viscosity ν_T is evaluated in a way similar to Prandtl's mixing length hypothesis,

$$\nu_T = l_m^2 |\overline{S}| \qquad\qquad l_m = C_s \Delta \qquad\qquad (2.20)$$

in which C_s is the Smagorinsky constant. This gives the final expression for the SGS stress as:

$$\tau_{ij}^{R} = -2\Delta^2 C_s^2 |\overline{S}|\overline{S_{ij}} \qquad\qquad (2.21)$$

Thus the Smagorinsky model implies that the SGS stress tensor and the filtered strain rate tensor are aligned and can be related through a single constant C_s. We note that no explicit filtering is needed to implement the SGS model. In our code variable values on the grid are taken as filtered values, which implies the application of a box filter with a (variable) width equal to the cell size. Thus, it is not possible for us to determine an exact filter function in order to compare our results with DNS, as would be the case with any other complex geometrical simulations. We also note that the turbulence production term, which is the inner product of the SGS stress τ_{ij}^{R} and the filtered strain $\overline{S_{ij}}$, is negative definite implying that energy is being transferred from the large scales to the small scales. This is only qualitatively correct, and it does not allow reverse energy cascades or backscatter. By studying the behavior of the model in the inertial range various authors have made predictions to estimate the constant C_s. Lilly (1967) first predicted a value of 0.17; others have predicted lower values ranging from 0.065 to 0.1. Unfortunately there is no common agreement on the value of C_s which is determined empirically. The more complicated the flow gets the more difficult it is to predict the model constant C_s, and no such value is known for separated shear flows with curved streamlines, as in our case. In our simulation we use Cs = 0.1, as predicted by Piomelli et al. (1988).

In addition to the ambiguity of C_s the Smagorinsky model has many drawbacks. In most commercial codes the model must rely on ad hoc methods to extrapolate sub grid scale (SGS) shear stresses near the wall. Our CFD-ACE code uses the well known Van Driest (1956) damping function to locally extrapolate eddy viscosity to the wall. The behavior of the model at the wall is especially important to our simulation since we calculate shear stress at the wall And finally, since the model constant is fixed, the model does not allow energy flow from small scales to large scales which can be significant (Germano et al. 1991) and hence produces excessive disspation of large scale fluctuations.

2.5.2 The dynamic model

The dynamic model, originally due to Germano et al. (1991), is also an algebraic SGS model. Here, in addition to the subgrid filtering, another filter called the subtest filter is applied to the flow field. Typically the width of the subtest filter is chosen to be twice the width of the subgrid filter. Our code uses implicit filtering for the subgrid level and explicit filtering with a top-hat filter (in all three directions) for the subtest level. We denote the subgrid filtering with an overbar and the subtest filtering with a tilde. Then, using the eddy viscosity hypothesis and a Smagorinsky-type model for the subgrid and subtest stresses, we obtain,

$$\tau_{ij}^R = (\overline{u_i u_j} - \overline{u_i}\ \overline{u_j}) = -2\overline{\Delta}^2 C |\overline{S}| \overline{S_{ij}} \tag{2.22}$$

$$\mathbb{T}_{ij} = \left(\widetilde{\overline{u_i u_j}} - \widetilde{\overline{u_i}}\ \widetilde{\overline{u_j}} \right) = -2\widetilde{\overline{\Delta}}^2 C |\widetilde{\overline{S}}| \widetilde{\overline{S_{ij}}} \tag{2.23}$$

where we denote the subgrid scale stress by τ_{ij}^R and the subtest level stress by $\mathbb{T}_{ij}$. Here we have replaced the C_s^2 (in Eqn. 2.21) by C to allow for the variation of sign. It is easy to see that the Leonard stress tensor defined by,

$$\mathbb{L}_{ij} = \mathbb{T}_{ij} - \widetilde{\tau_{ij}^R} = \widetilde{\overline{u_i u_j}} - \widetilde{\overline{u_i}}\ \widetilde{\overline{u_j}} \tag{2.24}$$

is a known quantity, and it can be used to evaluate the model constant. The Leonard stress tensor may also be written as,

$$\mathbb{L}_{ij} = -2C \left[\widetilde{\overline{\Delta}}^2 |\widetilde{\overline{S}}| \widetilde{\overline{S_{ij}}} - \overline{\Delta}^2 \widetilde{|\overline{S}| \overline{S_{ij}}} \right] \tag{2.25}$$

This equation may be used to evaluate C, but a single constant C is needed from the 5 independent components of the anisotropic part of L. To overcome this Lilly (1992) minimized the error using a least square technique. This procedure, however, leads to numerical instabilities, hence most implementations average the coefficient in the homogeneous direction, as proposed by Piomelli (1993).

There are several advantages of using the dynamic model compared to the Smagorinsky model. Firstly, the model coefficient is neither prescribed nor remains constant, rather it is determined as a part of the solution. Secondly, the Leonard tensor is zero in laminar flow, giving the correct zero SGS stress. Thirdly, the model predicts a cubic behavior of the SGS stress near the wall, which agrees well with experimental results. Also, the model can do away with *ad hoc* modifications to the SGS near the wall, as is commonly done in the Smagorinsky model. Lastly, the model constant C can take negative values, and hence the model can account for energy transfer in both directions.

2.5.3 The localized dynamic model

The localized dynamic model first proposed by Menon and Kim (1997) is a one-equation SGS model based on a method of first solving a *model* transport equation for the subgrid scale kinetic energy k.

$$k = \frac{1}{2}\left(\overline{u_i^2} - \overline{u_i}^2\right) \tag{2.26}$$

$$\frac{\partial k}{\partial t} + \overline{u_i}\frac{\partial k}{\partial x_i} = -\tau_{ij}^R\frac{\partial \overline{u_i}}{\partial x_j} - \epsilon\frac{\partial}{\partial x_i}\left(\nu_T\frac{\partial k}{\partial x_i}\right) \tag{2.27}$$

The three terms on the right hand side of Eqn. 2.27 represent the production, dissipation and transport of SGS kinetic energy. Here the SGS stress is modeled using the eddy viscosity hypothesis, the eddy viscosity is modeled using the SGS kinetic energy and dissipation is also modeled using the SGS kinetic energy on dimensional grounds. This procedure is similar to that used in the one-equation Reynolds Averaged methods.

$$\tau_{ij}^R = -2\nu_T\overline{S_{ij}} \qquad \nu_T = c_\nu k^{1/2}\overline{\Delta} \qquad \epsilon = c_\epsilon\frac{k^{1/2}}{\overline{\Delta}} \tag{2.28}$$

The model constants c_ν and c_ϵ are evaluated by applying the dynamic modeling method (as described above) to the kinetic energy equation. This SGS model removes the mathematical inconsistency of the algebraic dynamic model (having to approximate one constant from five equations), and because the model computes the evolution of SGS kinetic energy, it is capable of capturing non-local and history effects of the turbulence. This is the central advantage of the model over other algebraic models.

2.6 Structural modeling

To compute the response of the arm-suspension structure obstructing the flow we employ a finite element stress solver module included in the CFD-ACE code that can be directly coupled to the flow solver. Equations of structural mechanics are solved in finite element form, derived from the principal of virtual work. For each element displacements are defined at the nodes and obtained within the element in the usual manner, by interpolation from the nodal values using the shape functions.

2.6.1 Structural damping

To treat structural damping in the actuator arm we use the simple Rayleigh damping (proportional damping) method. By allowing,

$$\mathbf{C} = \alpha\mathbf{M} + \beta\mathbf{K} \tag{2.29}$$

for each degree of freedom of the structural model,

$$\xi_i = \frac{1}{2}\left(\frac{\alpha}{\omega_i} + \beta\omega_i\right) \tag{2.30}$$

where ξ is the damping ratio and ω in the natural frequency. Such a formulation (Cook et al. 1989) permits us to choose the amount of damping for two frequencies. In our particular case we choose 2% of critical damping at the first and tenth modes of vibration of the structure. This effectively guarantees that the damping in the spectrum of interest (first ten modes, 1-40 kHz) is below 2% and vibration modes outside this range are heavily damped out.

2.6.2 Coupling of the fluid and structural models

On the completion of one time-step by the flow solver, pressure data (normal loading) and shear stress at the wall (tangential loading) are passed on to the stress solver. These forcing boundary conditions are implemented on a face-by-face basis, without the use of simplifying assumptions. The resultant forces on the actuator may be obtained by integrating the pressure and shear loads over the surface area of the actuator:

$$\int_{\partial S} p n_i dS \qquad \int_{\partial S} \left(\tau_{ij} + \tau_{ij}^R \right) n_j dS \qquad (2.31)$$

Theoretically the SGS stress τ_{ij}^R should asymptote to zero at the walls, however, this is not always the case practically, hence it is included in the integration of the shear stress. The FE stress solver then determines the response of the structure as the simulation progresses. The deflections of the actuator arm (of the order of a few hundred nanometers) are usually very small compared to the grid size in the vicinity of the arm. Such deformations are also small compared to the mean free path of the fluid (65 nanometers), hence, there is no need to feed back the structural solution to the flow solver. Thus all our simulations are unidirectionally coupled from the flow to the structure.

2.7 Model setup

Unlike in experimentation, computational investigations provide relative ease in simulating different geometries and configurations. This book includes simulations for various geometries, form factors and arm positions. In this Chapter two basic simulation models are described. In later Chapters, when different models and constructed and simulated, they are described individually as and when necessary.

In this Chapter, two generic disk drive models and built and simulated. Both models, shown in Figures 2.1 and 2.2, consists of two 3.5" disks rotating at 10,000 RPM in a fixed, closed enclosure. The difference between the models lies in the details of the actuator that is placed symmetrically between the rotating disks. Figures 2.1 and 2.2 also show an exaggerated schematic of the cross section between the disks. The first model consists of a single e-block arm only, while the second actuator consists of a more realistic actuator, containing the suspension and slider unit. Table 2.1 gives material properties for different components of the actuator. Table 2.2 provides details of the geometric dimensions of each component in the models.

The grids that were used in the large eddy simulations are shown in Figures 2.3 and 2.4. Generally since the dimension in the $r - \phi$ plane is much larger than the in z direction, the grid is generated by first entirely specifying the grid in the two in-plane dimensions and then it is extruded along z. Hence, all computational cells are hexahedrals and we thus avoid the inaccuracy and numerical dissipation of tetrahedral volumes. The azimuthal symmetry of the shrouded portion of the drive (for a span of 200-250 degrees) allows the use of a block structured grid in that region. In the region close to the arm the grid is unstructured to conform to the complicated geometry of the actuator arm. However, instead of using the traditional triangular elements we use unstructured quadrilaterals which typically require lesser cells for the same average resolution. Generally with elements that accurately conform to the boundaries our unstructured grids contain 5-10% triangular elements, while the rest are quadrilateral elements. In Chapter 5 the grid generation is discussed in more detail, and the uncertainty due to the grid itself is quantified.

Figure 2.5 shows the location of the shroud wall (which forms the external boundary of the domain) with respect to the disk. The disk to shroud clearance is 1 mm. Actuators that are used in current disk drives are made up of certain specialized structural components. Figure 2.6 shows a close up of the actuator used in Figure 2.2 and also shows the nomenclature used for each component. Figure 2.6 also shows two directional arrows labeled "off-track" and "on-track". These schematic arrows have been shown to depict the definitions of off-track and on-track that have been used in the rest of this book. The off-track direction is taken to be perpendicular to the longitudinal axis of the actuator, while the on-track direction in taken to be parallel to this axis. The model shown in Figure 2.1 contains an e-block arm only, and a close up of the arm is shown in Figure 2.7.

It is important to realize the structures are three dimensional and the axial dimension of the structure is modeled and simulated in all our simulations. A three dimensional view of the e-block arm used [2] in Figure 2.1 and the entire actuator used in Figure 2.2 is shown in Figures 2.7 and 2.8 respectively.

2.7.1   Boundary and Initial Conditions

The boundary conditions for the computational domain are shown in Figure 2.9. They are implemented as follows:

- The top and bottom disks (along with the central hub) are modeled as rigid rotating walls. Effects such as run-out (especially non-repeatable run-out, NRRO), clamping distortions and disk vibrations cannot be accounted for in this model.

- The actuator (either the e-block arm or the complete actuator) is modeled as a fixed obstruction to the flow, with no-slip boundary conditions. The back face of the actuator, flat in the case of Figure 2.1 and curved in the case of Figure 2.2 is fixed. Generally one of the causes of off-track motions is the rigid body motion of the actuator, due to slip at the actuator bearing. This slip (sometimes referred to as the

---

[2] The term "e-block" derives from arm structures to which three or more arms could be attached

actuator buffetting) is not accounted for in our simulations. The slider is modeled as being simply supported on the disks, i.e. the small fly-height head-disc spacing with the inherent air bearing stiffness is not modeled. There are no computational cells representing the air bearing and the slider is free to slide on the rotating disk.

– In the computational domain the shroud gaps are modeled as symmetric boundaries (slip wall boundary conditions). This ensures that in the gap, airflow is permitted only in the plane of the disks, but not perpendicular to them. Since the addition of cells to the top and bottom of the current domain is not computationally feasible, this is a good approximation to the narrow shroud gap. Alternatively, a periodic boundary condition may be enforced between the top and bottom shroud gaps, such that the flow leaving the domain at the top re-enters the domain at the bottom. This boundary condition, however, led to unphysical travelling waves in the velocity solutions, and hence was not used.

The LES simulation is initialized from a steady state RANS (Reynolds Averaged Navier Stokes) solution, using the standard $k - \epsilon$ model. To this solution, we add 5% random fluctuations to velocities, to perturb the base flow. This implies that for the initial condition, the flow field is assumed to be in steady state with the rotating disks, with small perturbations from the mean. In Chapter 5, Section 5.2.5, a detailed discussion provides the justification for the use of such initial conditions and also estimates the dependence of our solutions on the initial conditions.

2.8 Results

2.8.1 Modal Analysis

To begin, a modal analysis is performed to estimate the natural frequencies of vibration of the single e-block arm and the entire actuator. The natural frequencies and mode shapes are listed in Tables 2.3 and 2.4 respectively. The first four modes of the e-block arm are also shown in Figures 2.10–2.13. Of particular interest to the current problem are the sway modes, which contribute most to the off-track displacement, while the lower bending modes contribute primarily to the on-track displacement. All of the results presented in this Chapter refer to the simulations using the Dynamic LES model. The three different SGS models described above are tested in the next Chapter.

2.8.2 Flow Field

The flow inside a disk drive casing is very complex and involves regions of mostly transitional and turbulent flows. For the configuration in our simulations the flow field near the center of the rotating disks remained transitional, while most of the other regions – including the sheared region at the shroud and the wake – was largely turbulent.

The flow displays a strong stagnation zone near the leading edge of the arm. The top and bottom surfaces of the arm contain regions of unsteady separation and reattachment which results in the formation of coherent structures, particularly in the wake of the arm. The wake itself is very complex showing regions of shear and the presence of intense vortices that are continually being transported due to the shearing effect of the rotating walls at the top and bottom. Due to the lack of symmetry of the arm with respect to the incoming flow there appears to be no strict periodicity in the shedding of vortices.

Figure 2.14 and Figure2.15 show the time averaged contour plots of the azimuthal and radial velocities, respectively, in the mid plane of the model, averaged after 5 revolutions of the disk, for the first simulation model. The figures show a largely uniform flow field in about the 3/4 portion of the drive upstream of the arm. The wake region contains a more irregular flow topology. Interestingly there is a small region of flow reversal, near the hub, just upstream of the arm. This flow reversal is probably due to the adverse pressure gradient (the flow stagnates at the arm). The radial contours show a strong inflow in the wake of the arm; this is primarily due to the constraining geometry and the disk rotation.

There are several methods to quantify the "turbulence level" of a flow. Some of them include: turbulence intensity (used in this Chapter, and Chapters 3,5 and 8), Magnitudes of Reynolds stresses (used in Chapter 4), RMS of velocities (used in Chapters 3, 5, 6, 7 and 8), Energy Spectra (used in Chapters 3 and 5) or the contribution to RMS fluctuations from different frequency bands, (used in Chapters 5, 6 and 7). Higher order statistics of the flow (such as 3rd and 4th moments of velocity or kinetic energy) and intermittency are not used in this book.

Figure 2.17 shows the values of turbulence intensity along chords 1-5 that are defined in Figure 2.16. The turbulence intensity (sometimes expressed as a percentage) is defined as the ratio of the RMS of the velocity and the mean velocity.

The plots in Figure 2.17 demonstrate the high levels of turbulence generated in the wake formed behind the arm for the model in Figure 2.1. As the location of the chord increases in its azimuthal distance from the arm, the turbulence intensity decreases, mainly due to viscous and sub-grid scale dissipation. The bulk of the turbulent fluctuations shift towards the center radius of the disks due to the constricting nature of the shroud wall. As the flow comes back to approach the arm the turbulence does not dissipate completely, but is approximately 20% of the turbulence intensity in the wake.

The velocity profile of the flow between the two disks is very similar to a turbulent Couette flow. Figure 2.18 shows the inter-disk velocity profile at four azimuthal locations for the model in Figure 2.2. The velocity is projected into it's radial and azimuthal components. The azimuthal velocity shows several interesting features: the black line shows the profile immediately in the wake of the arm. Here the velocity has the "smallest profile", indicating that the flow has lost some momentum due to the formation of the turbulent wake. As the flow moves on towards subsequently increasing azimuthal positions, the flow gains momentum by convection/diffusion from the rotating disks and the velocity profile becomes "fuller". The thickness of the boundary layer increases with the azimuthal position. On the other hand, the radial velocity in Figure 2.18 is highly negative in the wake – indicating a strong radial inflow downstream of the arm. This is mainly due to the constricting geometry of the shroud. As the presence of the shroud reduces the radial pressure gradient, the flow shows smaller radial velocities as seen in Figure 2.18. The positive peaks in radial velocity that occur near the disks in both Figures 2.18 and 2.19 are due to the centrifugal effect of the rotating disks.

Figure 2.19 basically shows that the mean velocity upstream of the arm attains solid body rotation characteristics. This is evident by the linear dependence of the azimuthal velocity with radius. Also, the velocity profiles shown in the Figure 2.19 very closely resemble those of turbulent Couette flow.

2.8.3 Vortex dynamics

Figure 2.20 shows a three dimensional view of instantenous streamtubes in the region of the wake behind the arm. They are color-coded (for contrast) according to the azimuthal velocity. Figure 2.21 shows the orientation of these tubes with respect to the midplane. Additionally the midplane is colored to reflect the axial velocity of the flow. These figures demonstrate the orientation of vortical structures shed by the arm, whose axes are oriented at an angle to the arm. This is most likely due to the "forcing" of the disks, and results in an orientation that is different compared to the wake of a regular cylinder. Eddies are typically generated from every sharp edge of the arm and transported downstream by the Couette type flow. These structures are coherent and persistent; viscous dissipation does not cause them to be dissipated completely before they approach the arm after being transported to its upstream side.

2.8.4 Drag

Figure 2.22 and Figure 2.23 show the pressure drag and the viscous drag on the e-block arm as a function of time. We define drag as the net resultant force acting in a direction perpendicular to the axis of symmetry of the arm. These have been obtained by integrating the pressure and shear stress on the area of the arm using Eqn. 2.31. The area ∂S includes all of the surfaces of the arm (including the surfaces formed by the holes). We note that the viscous drag (or the skin friction drag) is two orders of magnitude smaller than the pressure drag, and hence the corresponding contribution of the pressure drag to the vibration of the arm is significantly higher.

Figure 2.24 shows the frequency spectrum of the total drag. We see that the power of the spectrum is concentrated in the low frequency (0 - 4 kHz) range, and the higher frequency part of the spectrum is more uniform. This implies that we can (numerically) expect a low frequency forcing of the e-block arm by the flow. Moreover, none of the velocity or pressure frequency spectra show sharp peaks at a single well defined frequency. This leads us to believe that the flow-structure interaction does not get locked into a Strouhal frequency and the vortex shedding process is highly unsteady and random. Frequency spectra are discussed in more detail in Chapters 5-8.

2.8.5   Spatial Variation of Pressure

Given that pressure contributes the most to the vibrations of the e-block arm, we discuss the pressure fluctuations in the flow field along the leading edge face of the arm. In particular, we note the pressure at ten points on the leading edge face of the arm as shown by the small dots in Figure 2.25. The points are numbered so that they start from 1 at the tip of the arm, and go to 10 at the fixed pivot of the arm. Figure 2.26 shows a waterfall plot, where each line denotes the frequency spectrum of pressure fluctuation at that point. From this figure we again note that the pressure fluctuations are rich in the low frequency range. It is of interest to note that point #9 displays a significantly higher amplitude (of the spectrum) in the low frequency range than its neighbors. This is most likely due to the fact that the upstream incoming velocity is the highest at this point, and this results in a large pressure rise as the flow stagnates at the face of the arm.

2.8.6   Windage

Finally, we also calculate the "windage" loss at the disks, shown in Figure 2.27. This quantity refers to the power required by the motor to rotate the disks at 10,000 RPM due to viscous effects. In general there is no agreement on the definition of the term "windage". Some authors (Tsuda et al. (2003)) use the term to imply the disk power loss (in watts), while others use it more generally to refer to "the fluctuating aerodynamic force" (Shimizu et al. (2003)) and some others (Hirono et al. (2004)) use "windage" to refer to the flow-induced displacements of the arm. We prefer to use windage to refer strictly to the power loss at the rotating disks due to viscous action. This quantity may be calculated using the expression. Windage may be easily calculated as,

$$W = \int_{\mathscr{A}_d} \left(2\nu \overline{u_j} \overline{S_{ij}} - \overline{u_j} \tau_{ij}^R \right) d\mathscr{A} \tag{2.32}$$

where W is the windage and $\mathscr{A}_d$ is the area of the disks. The calculated value of windage in watts agrees very well the experimental and computational results of Tatewaki et al. (2001). Figure 2.27 shows a time history of the windage loss. We note that this estimate of windage considers only 1 face of each of the 2 rotating disks. In an actual drive windage is due to power lost on both faces of each rotating disk.

2.8.7   Vibrational response of the arm

The vibrations of the structures as a response to the flow described above are reported in terms of off-track and on-track displacements.

For the first model with a single e-block arm, Figure 2.28 and Figure 2.29 show the displacements of the end of the e-block arm in the off-track and on-track directions respectively. Figure 2.30 and Figure 2.31 show the corresponding frequency spectra. From the figures, we conclude that the off-track amplitude is limited to about 2.2 nm peak-to-peak, with a mean at about 2.5 nm. The on-track vibration is significantly greater, with a peak to peak amplitude of about 5.2 nm, with a mean at about 2 nm. We conclude that the response of the arm in the on-track direction is larger due to its lower stiffness in bending. Since the arm is modelled here as a cantilever its lowest stiffness is in bending, and this causes relatively large bending vibrations (out of plane vibrations) as shown in Figure 2.32. Large on-track displacements are simply a consequence of the bending.

The frequency spectra of the vibrations correlate very well with the modal analysis. In Figure 2.30 peaks are seen at 5.785 kHz (very close to the second bending mode), 7.621 kHz (first torsional mode) and a large peak at 8.901 kHz (close to the first sway mode). In the on-track spectrum, a large portion of the power is concentrated in the region close to the 1.252 kHz first bending mode, implying that the dominating frequency of oscillation corresponds to the first bending mode. Additionally, the second bending, first torsion and first sway modes are also evident.

For more physical insight into the vibrations we plot the trajectory of the point under consideration on the x-y plane in Figure 2.33, where the large dot represents its original undeflected position.

Generally we note that the vibrations of the e-block arm are primarily dominated by the first bending mode. In practice the boundary conditions for the drive level components are different, given that a suspension and slider is attached to the end of the arm. This may be studied using the simulation containing the full actuator model.

Figures 2.34 and 2.35 show the time history and frequency spectra of the vibrations of the complete actuator arm. The off-track and on-track vibrations are reported at the center of the slider while the out-of-plane motions are reported at the edge of the suspension. In this case, the peak-to-peak off-track vibrations of the slider are approximately 4 nm and on-track vibrations are approximately 5 nm. We also notice that the off-track mean is quite large, approximately 10 nm. The off-track deflection of the slider achieves its steady mean at about 0.012 s which corresponds to 2 rotations of the disk at 10,000 RPM.

The frequency spectra also correlate very well with the modal analysis in Table 2.4. The modes are: the first sway and the first torsion in the off-track direction, the first and second bending in the on-track direction and the first and second bending and the first torsion in the out-of-plane direction.

2.9 Conclusions

This Chapter seeks to provide an introduction to Large Eddy Simulation as a useful tool for studying flows in disk drives. Using LES our simulations provide rich data in terms of pressure and velocities. We have also been successful in integrating the flow and stress solvers, and the structural response results agree quite well with the modal analysis.

In terms of the flow topology we observe a highly complicated shear flow with aperiodic vortex shedding in the wake of the arm. The turbulent eddies are not dissipated completely by the time they complete one revolution and the upstream turbulence intensity is 10-15% of the intensity in the wake. The pressure fluctuations are rich in the low frequency (0-3 kHz) range and act as low frequency excitations to the structures. The pressure drag on the arm is two orders in magnitude larger than the viscous drag.

In terms of the response of the arm to the flow – the structure vibrates at frequencies corresponding to its first few modes. Due to the nature of the model (i.e. the arm being modeled as a cantilever) the arm shows relatively large vibrations in bending, which in turn contribute to the on-track displacement. The vibrations that are more important to designers, (i.e. off-track) correspond closely to the first sway mode, which in the case of the e-block arm has a frequency of 9.3 kHz. Similar modal excitations are seen in the case with the complete actuator. The off-track mean is shifted to 10 nm while the peak-to-peak is about 4 nm at the slider.

While this Chapter seeks to provide numerical results for the flow variables (velocities, intensities and pressure) and structural response (displacements and spectra), very little attention has been devoted to the role of the turbulence model, the internal numerics of the code or the grid used in the simulation. These shall be the topics of investigation in Chapters 3, 4 and 5.

In future Chapters we do not compute the flow induced vibration results of the actuator, for two reasons. Firstly, CFD-ACE (as of 2006) does not allow the computation of structural vibrations when performing simulations in parallel. The simulations presented in this Chapter are smaller in size (i.e. number of cells) and are able to be computed on a single desktop computer. Most simulations presented later on are much larger cases that need several CPUs. As a response to this constraint one may suggest the coupling of CFD and structural codes outside the setting of the commercial code CFD-ACE. However, off-track and on-track vibrations of the slider are dependant on several other factors beyond just the flow (as listed in Section 1.2) and modeling all those factors is beyond the scope of this book – this is the second reason. In Chapter 9, with the help of a valuable code developed by another researcher, we couple our CFD forcing data with a realistic model for the suspension elasticity and slider-disk dynamics.

2.10 Tables

Table 2.1: Material properties of actuator

e-block arm	Young's Modulus	69 GPa
	Density	2710 kg/m^3
	Poisson's Ratio	0.33
Base Plate	Young's Modulus	210 GPa
and Suspension	Density	8700 kg/m^3
	Poisson's Ratio	0.3
Slider	Young's Modulus	410 GPa
	Density	4350 kg/m^3
	Poisson's Ratio	0.3

Table 2.2: Geometry data

	Simulation 1	Simulation 2
Number of disks	2	←
Number of e-block arms	1	1
Number of base plates	0	2
Number of suspensions	0	2
Number of sliders	0	2
Disk thickness (mm)	1	←
Disk diameter (mm)	76.2	←
Width of shroud gap (mm)	1	←
Length of actuator (mm)	45	←
Length of e-block arm (mm)	32.5	←
Length of base plate (mm)	6.5	←
Length of suspension (mm)	11.1	←
Thickness of e-block arm (mm)	0.8	←
Thickness of base plate (mm)	0.3	←
Thickness of suspension (mm)	0.1	←
Dimensions of slider (mm)	$1 \times 0.8 \times 0.3$	←
Number of weight saving holes in e-block arm	2	←

Table 2.3: Natural Frequencies and mode shapes of the e-block arm

Mode Number	Natural Frequency (kHz)	Mode Type
1	1.252	First Bending
2	5.529	Second Bending
3	7.768	First Torsion
4	9.387	First Sway
5	13.792	Third Bending
6	16.877	Second Torsion
7	24.398	Fourth Bending
8	25.292	Second Sway
9	28.103	Third Torsion
10	40.733	Third Sway

Table 2.4: Natural Frequencies and mode shapes of the complete actuator

Mode Number	Natural Frequency (kHz)	Mode Type
1	1.417861	First Bending
2	4.784891	Second Bending
3	5.534802	First Sway
4	6.26456	First Torsion
5	10.77796	Third Bending
6	11.4283	Fourth Bending (Suspension)
7	11.75965	Second Sway
8	14.20358	Second Torsion
9	15.98683	Fifth Bending
10	21.72979	Third Sway (asymmetric)

2.11 Figures

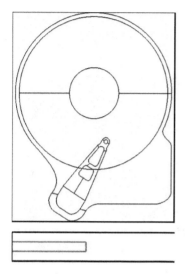

Figure 2.1: Front view and schematic sectional view of CFD model containing a single e-block arm only

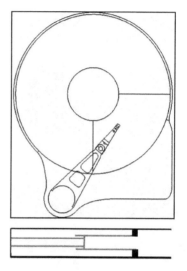

Figure 2.2: Front view and schematic sectional view of CFD model containing an e-block arm, base plates, suspensions and sliders

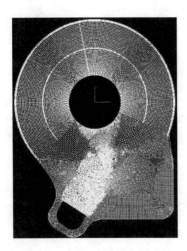

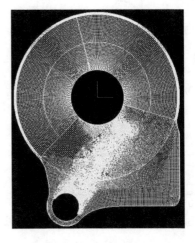

Figure 2.3: Plan view of hexahedral grid used to simulate geometry shown in Figure 2.1. The grid is block-structured in the symmetry region and grid density is increased upstream and downstream of the e-block arm

Figure 2.4: Plan view of hexahedral grid used to simulate geometry shown in Figure 2.2. The grid is block-structured in the symmetric region and grid density is increased upstream and downstream of the actuator arm

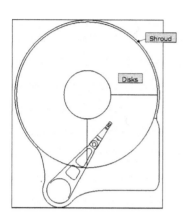

Figure 2.5: Location of the shroud with respect to the rotating disks

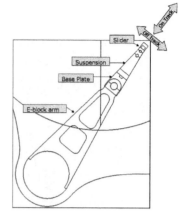

Figure 2.6: Closeup and nomenclature of the actuator used in typical disk drives

Figure 2.7: Three-dimensional view of e-block arm

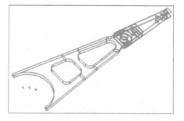

Figure 2.8: Three-dimensional view of actuator, showing the single e-block arm, two base plates, two suspensions and two sliders

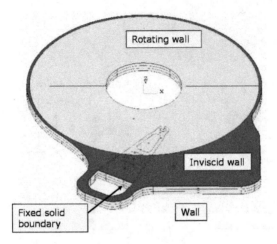

Figure 2.9: Overview of the boundary conditions used in the simulations

Figure 2.10: Mode 1: First Bending, 1.252 kHz

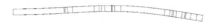

Figure 2.11: Mode 2: Second Bending, 5.529 kHz

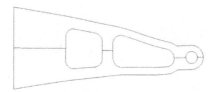

Figure 2.12: Mode 3: First Sway, 7.768 kHz

Figure 2.13: Mode 4: First Torsion, 9.387 kHz

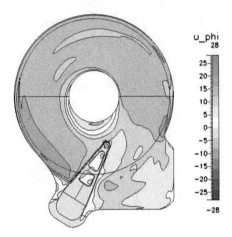

Figure 2.14: Time averaged contours of azimuthal velocity

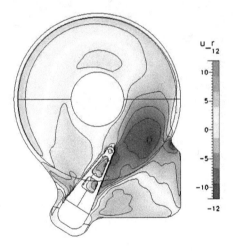

Figure 2.15: Time averaged contours of radial velocity

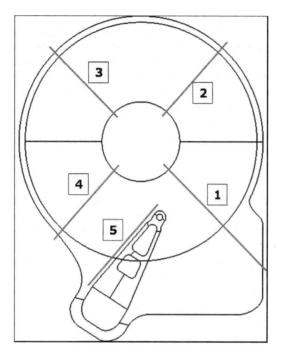

Figure 2.16: Locations of chords 1-5 on which turbulence intensity is plotted in Figure 2.17.

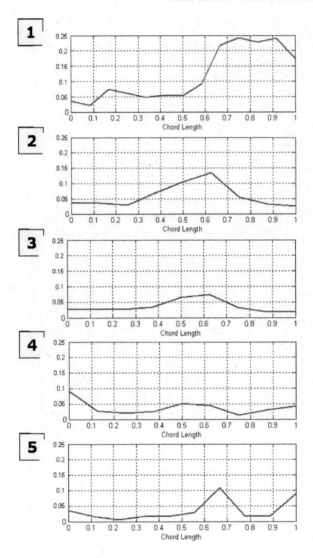

Figure 2.17: Turbulence intensity along chords 1-5 shown in Figure 2.16.

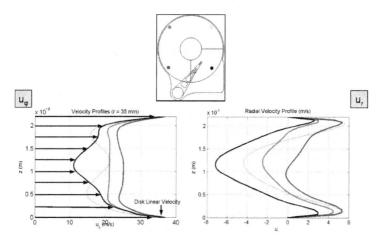

Figure 2.18: Profiles of the mean radial and azimuthal flow velocity as a function of azimuthal positions.

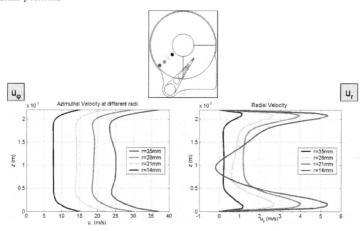

Figure 2.19: Profiles of the mean radial and azimuthal flow velocity as a function of radial positions.

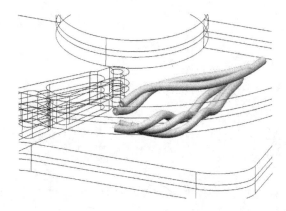

Figure 2.20: Three dimensional view of streamtubes in the wake of the arm

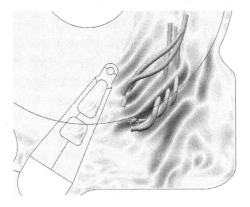

Figure 2.21: Orientation of streamtubes relative to the midplane of the model

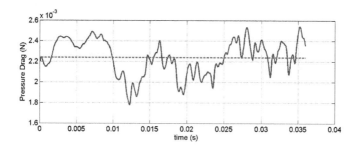

Figure 2.22: Pressure Drag on the arm as a function of time. The dotted line indicates the mean value

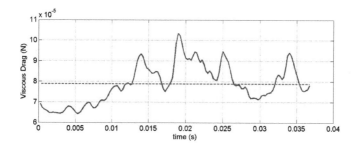

Figure 2.23: Viscous Drag on the arm as a function of time. The dotted line indicates the mean value

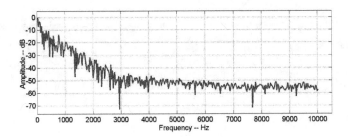

Figure 2.24: Frequency Spectrum of the Drag Force.

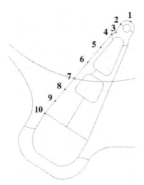

Figure 2.25: Schematic of points where pressure fluctuations are reported.

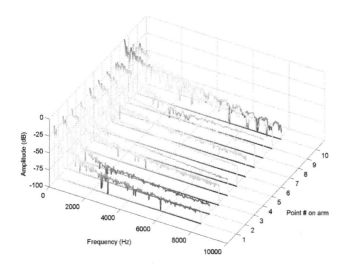

Figure 2.26: Waterfall plot showing the frequency content of pressure fluctuations at 10 points along the face of the arm.

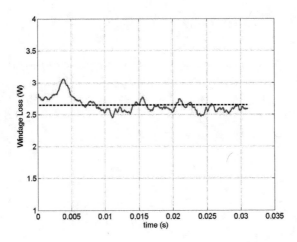

Figure 2.27: Windage loss at disks as a function of time.

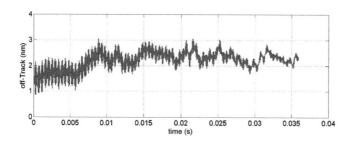

Figure 2.28: Off-track deflection of arm-tip (nanometers).

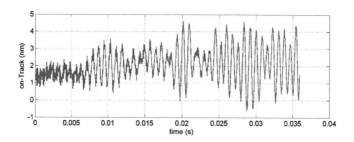

Figure 2.29: On-track deflection of arm-tip (nanometers).

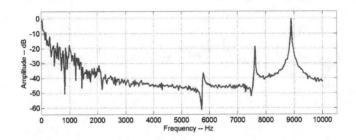

Figure 2.30: Frequency Spectrum of off-track deflections.

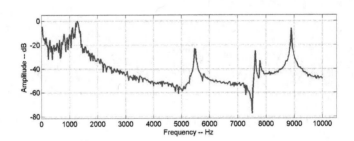

Figure 2.31: Frequency Spectrum of on-track deflections.

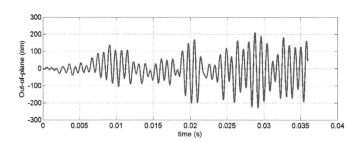

Figure 2.32: Out of plane (bending) deflections (nanometers).

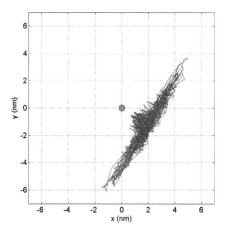

Figure 2.33: Plot of trajectory of the arm tip in the horizontal plane.

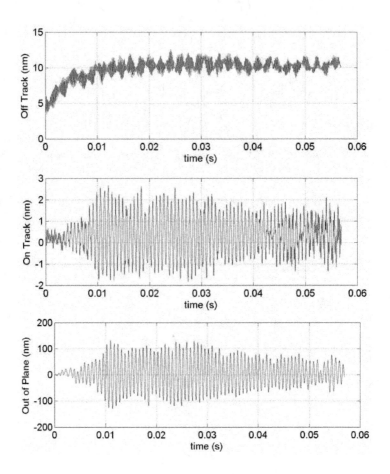

Figure 2.34: Off-track, On-track and out-of-plane flow indiced vibrations of the actuator. The off-track and on-track are reported at the slider, while the out-of-plane are at the suspension edge.

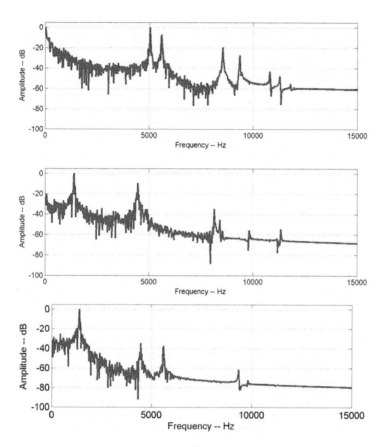

Figure 2.35: Frequency spectra of Off-track, On-track and out-of-plane vibrations of the actuator

Chapter 3

A Comparison of SGS Models

3.1 Introduction

As we saw in the last Chapter the air flow generated due to high speed rotating disks in modern computer disk drives is complicated and contains a range of attributes that require careful attention in a simulation. This Chapter presents an in-depth comparison of LES models, with an emphasis on accurate simulation of airflows in disk drives.

We also observed in the last Chapter that the flow field upstream of the arm (after one complete turn around), has a turbulence intensity of nearly 10-15%. Any change in the upstream turbulence will lead to changes in the pressure fluctuations at the e-block arm, corresponding to a different structural excitation. For this reason, it is important to model the turbulence dissipation (by subgrid transfer and by viscous action) correctly, i.e. the numerical differencing method and sub-grid scale turbulence model should be relatively free of artificial dissipation.

The flow field is also characterized by separation and vortex shedding at the trailing edges of blunt bodies in the flow. This random unsteady shedding of vortices leads to random changes in circulation around these bodies, resulting in unsteady aerodynamic forces. The turbulence model should be able to capture the vortex shedding and the associated form drag.

In modern computer simulations of disk drive enclosures very little attention is paid to the turbulence model used. This is often because, from a user's perspective, the inclusion of a turbulence model in a fluid dynamics calculation can be done very easily in commercially available CFD codes. On the other hand, there is very little experience in the use of LES models for disk drive airflow simulations. Usually the LES model is chosen indiscriminately, often resulting in less than accurate results.

To build credibility into a set of results, it is customary to perform either *a-priori* or *a-posteriori tests*. In the former, experimental or DNS data can be filtered to observe the performance of the LES model and direct comparisons of the predicted SGS stresses can be made. In a-posteriori testing, statistics of computed LES solutions may be compared with those obtained by experiments or DNS. Unfortunately for flows in disk drives limited experimental data (Gross 2003; Barbier 2006) and no DNS data is currently available in the literature, which considerably limits the scope of this exploration. Therefore we are limited to comparing the performance of different LES models only, but this comparison leads to valuable insights about the behavior of these models. We are able to compare the flow fields using these different LES models and relate the properties of the field to the property of the model. In the Sections that follow we make a comparison between the Smagorinsky model (Section 2.5.1), the Dynamic model (Section 2.5.2), the localized dynamic model (Section 2.5.3) and a direct simulation on the same grid as the LES calculations.

Similar comparisons of LES models appear in other works, such as Vreman et al. (1997) and Fureby et al. (1997). Although these works deal with simple flows there is excellent qualitative agreement in the results. This book presents the first such comparison applied to the complicated flows in disk drives.

3.2   Model Setup

This Chapter uses the same computational model as used in Chapter 2, with the single e-block arm only, shown originally in Figure 2.1. The Figure is repeated in this Chapter, as Figure 3.1. The model consists of 2 disks, rotating at 10,000 RPM, separated from each other by a gap of 3 mm. The gap between the disk outer edge and the enclosing wall (shroud) is 1 mm. A single obstruction in the form of an e-block arm was used. The thickness of the arm is 1 mm, and it is placed symmetrically at the midplane between the disks. The horizontal boundary surfaces of the computational volume at the top and bottom (except the rotating disks), were modeled as an inviscid wall (symmetry plane boundary conditions). The structure was fixed at its back face and thus modeled as a cantilever. Each simulation was started from the same initial conditions, which were obtained from a steady state k-epsilon solution of the average flow.

An unstructured grid, with quadrilateral dominant cells (90% quadrilateral elements, 10% triangular elements) was used. The grid is shown in Figure 3.2. The total number of cells was 245,745, the smallest volume was 6.13e-12 m^3 and the largest volume was 5.41e-10 m^3. A representative grid size of 0.408 mm may be calculated by averaging over all the control volumes as,

$$h = \left(\frac{1}{N} \sum_N \Delta V \right)^{1/3} \tag{3.1}$$

It is important to compare our grid size with the Kolmogorov scale and the Taylor micro-scale. The Kolmogorov scale gives an estimate of the length scale at which dissipation takes place. Ideally, direct numerical simulations resolve the Kolmogorov scale and require no artificial SGS-type dissipation. Using the $k - \epsilon$ method we are able to approximate the dissipation, ϵ, in our computational volume. Dissipation is obviously a function of position, but when averaged over the entire domain it is found to be approximately 9.78×10^4. We note that this value is in good agreement with the dissipation predicted by the large eddy simulations (see Table 3.2). The upper bound on dissipation was 5.64×10^5. We used this average estimate of dissipation to approximate the Kolmogorov length scale, and the velocity and time scales,

$$\eta = \left(\frac{\nu^3}{\epsilon}\right)^{1/4} = 0.0143mm \tag{3.2}$$

$$u_\eta = (\epsilon\nu)^{1/4} = 1.126m/s \tag{3.3}$$

$$\tau_\eta = \left(\frac{\epsilon}{\nu}\right)^{1/2} = 1.267 \times 10^{-5}s \tag{3.4}$$

This calculation indicates that our grid size is one order larger than what is needed to resolve the Kolmogorov scale. Hence the direct simulation cannot represent all of the dissipating motions, and therefore we should observe a build up of excessive small scales. This prediction is later confirmed.

Another method to estimate a length scale is the Taylor's microscale, λ. Although it does not have a clear physical meaning (Pope (2003)), the Taylor scale may be used as an estimate of intermediate size eddies (at sufficiently high Reynolds numbers). For calculating the Taylor scale, the size of the largest eddies (L) is taken as the separation distance between the disks, i.e. 3 mm. To approximate the velocity scale of the largest eddies, we use 5% of the disk linear velocity to obtain,

$$\lambda = L\sqrt{10}Re_L^{-1/2} = 0.455mm \tag{3.5}$$

This calculation shows that our grid resolution is sufficient to resolve the Taylor scale λ. There is excellent agreement of the above length, time and velocity scales with the recent work of Kazemi (2004).

A time step of 2×10^{-5} was chosen, which allows us to resolve a frequency range up to 25 kHz. Given that the important dynamics of the flow is in the low frequency range as observed in the last Chapter, a 25kHz resolution is sufficient to resolve the essential physics of the flow. In order to compare the results from different turbulence models we used the same grid in each simulation. Each simulation was integrated for 2400 time steps, which at 10,000 RPM, corresponds to 8 revolutions of the disks. A conjugate gradient method was used to solve the elliptic Poisson equation for pressure (in the SIMPLE procedure) in all the simulations, and the over-relaxation parameters for each dependant variable (which control the speed of convergence) were the same.

3.3 Results and Discussion

3.3.1 Kinetic energy

We define the resolved kinetic energy as,

$$E_f = \frac{1}{2}\overline{u_i}\,\overline{u_i} \tag{3.6}$$

and a conservation equation for this quantity may be easily derived as:

$$\frac{DE_f}{Dt} - \frac{\partial}{\partial x_i}\left[\overline{u_j}\left(2\nu\overline{S_{ij}} - \tau_{ij}^R - \frac{\overline{p}}{\rho}\delta_{ij}\right)\right] = -2\nu\overline{S_{ij}S_{ij}} - \tau_{ij}^R\overline{S_{ij}} \tag{3.7}$$

The convention from the previous Chapter is continued, i.e. $\bar{\cdot}$ represents filtering, while $\langle\cdot\rangle$ represents time averaging. Eqn. 3.7 is examined in more detail in Chapter 5 in determining the role of the initial conditions.

When integrated over the entire volume the second term on the left hand side of Eqn. 3.7 should be zero, for a closed computational volume. Numerically, however, this term is not zero, and its value is a measure of the numerical dissipation of the simulation. The first term on the right of Eqn. 3.7 is the viscous dissipation (which is always negative by the second law of thermodynamics) and the second term is the loss of kinetic energy to the residual scales (i.e. production of residual kinetic energy k). This term is always negative for the Smagorinsky model, but it can change sign in the other two models. A positive SGS dissipation term implies the backscatter of energy from small scales to large ones. Direct simulations that do not calculate the SGS stress tensor τ_{ij}^R have zero SGS dissipation. Finally, the terms $\left(2\nu\overline{u_j S_{ij}} - \overline{u_j}\tau_{ij}^R\right)$ are a source term for the kinetic energy, which represents the work done by the rotating disks on the fluid volume. This rate of energy input is equal to the power loss at the disks, which we refer to as windage.

3.3.2   Windage

Plotted in Figure 3.3 is the resolved kinetic energy integrated over the entire volume, and Figure 3.4 shows the windage. A legend for all of the plots in this Chapter is given in Table 3.1. We note that there is very little difference in the global kinetic energy between the different turbulence models. This implies that the main features of the resolved flow are reasonably independent of the SGS model. This is in good agreement with Fureby et al. (1997). However, the direct simulation predicts approximately 1-5% less kinetic energy. On the other hand, as seen in Figure 3.4, the energy input into the system (i.e. the windage) is about 20% more for the dynamic model and localized dynamic model, and about 15% more for the Smagorinsky model, as compared to the direct simulation. This indicates that there are significant differences in the energy transfer mechanisms of these simulations. Clearly, the direct simulation has the least amount of energy input from the disks (W), but it exhibits kinetic energy comparable to the LES simulations. This is because the direct simulation lacks a mechanism to transfer energy to the unresolved scales (τ_{ij} is zero) which leads to the accumulation of too many small resolved scales. On the other hand, there is less than 1% difference in the kinetic energy between the dynamic model, the localized dynamic model and the Smagorinsky model, but an approximate 5% difference in their energy input rates. This indicates that although the Smagorinsky model has less energy input per unit time, it bears almost the same kinetic energy as the dynamic model. More insight into this discrepancy can be obtained by considering the way each model resolves the wall layer.

For practical considerations the first grid point from the wall in our simulations was maintained at $8 < y^+ < 20$. This ensures that the first grid point is between the viscous sublayer and the inertial sublayer. In the CFD-ACE code the Smagorinsky constant is damped near the wall using the well known van Driest damping function (Van Driest (1956)).

$$C_s = c_s \left(1 - e^{y^+/A}\right) \qquad (3.8)$$

where A is taken to be 26, as customary. The dynamic models do not use any damping functions and are known to display the correct asymptotic behavior at the wall (Germano et al. (1991)). This indicates that the Smagorinsky model's wall functions (which have no physical grounds, and are implemented only to agree with experimental results) are inaccurate in representing the velocity field close to the disks (hence the shear stress at the disks, and correspondingly the windage). Thus more confidence may be placed in the results due to the dynamic models, and we conclude that the Smagorinsky model is not accurate in representing the energy flow into the system, and this may have serious consequences on the physics of the flow.

Finally, we observe that the global kinetic energy of all simulations asymptote to the same value, which indicates that our simulations have a tendency to equilibrate to the same energy level although there exist differences between the amount of energy input per unit time. This is a surprising result, and it suggests that different energy production (windage) and dissipation (SGS dissipation, viscous dissipation and numerical dissipation) mechanisms have compensated each other. For this reason, the physics of these flows are reasonably similar in the mean.

3.3.3 Mean and RMS fluctuations of azimuthal velocity and pressure

The azimuthal and radial velocities can be decomposed into their mean and fluctuating components. Turbulence intensity is a non-dimensional quantity representing the ratio of the RMS of the fluctuation to the mean flow speed. We calculate the mean and RMS components of the azimuthal velocity, starting the averaging at 2 revolutions and ending it at 8 revolutions of the disk, i.e. averaging over 6 revolutions, or 1800 time steps.

We plot these turbulence statistics along four radial chords in the flow domain, each of them located midway between the disks (the chords used here are the same ones used in Chapter 2, Figure 2.16. The locations of the chords are shown again in Figure 3.5). The chord 1 lies in the turbulent wake formed behind the arm, and the chords 2, 3 and 4 are at successively increasing angular positions along the direction of rotation of the disk. Mean flow velocities are plotted for each chord in Figures 3.6 to 3.9. RMS values of the fluctuations are plotted in Figures 3.10 to 3.13. Finally, the ratio of the two, (sometimes also referred to as the turbulence intensity) are plotted in Figures 3.14 to 3.17. To generate these figures, we used data at 10 points (12 points for chord 1) along the chord and a shape preserving spline interpolant was fit through the points. For all chords, 0 represents the inner boundary at the hub, and 1 represents the other boundary at the shroud.

We note that the regions close to the disk hub in Figure 3.13 representing chord 4 should be neglected from this analysis. The mean flow speed is small near the hub (often the flow reverses direction) resulting in large turbulence intensity (which is calculated using the mean flow in the denominator). Such turbulence intensity values (of the order of 100%) are unphysical.

For chord 1 we see that there is better agreement in the mean velocities and the RMS fluctuations predicted by the direct simulation and the dynamic models than with those predicted by the Smagorinsky model. We observe that the Smagorinsky model predicts significantly smaller fluctuations, which is an indication of its diffusive nature.

The same observations apply to chords #2, 3 while at chord #4 the difference between the mean velocities predicted by the models becomes insignificant. From all the figures illustrating the RMS fluctuations of velocity we can conclude that the Smagorinsky model has a tendency to predict lower fluctuations than the other turbulence models. This is evidence of the well known fact that the Smagorinsky model is overly diffusive and delays the transition of laminar to turbulent flow.

The above analysis is also consistent with our global kinetic energy diagram in Figure 3.3. The smaller fluctuations of the Smagorinsky model do not affect the total kinetic energy very much – fluctuations which are 5% of the mean contribute only 0.25% to the total kinetic energy.

Finally, plotted in Figures 3.18 to 3.21 are the mean pressures along the chords 1-4 and in Figures 3.22 to 3.25 the RMS values of pressure fluctuation are plotted along chords 1-4. Very little variation in the mean pressure is consistently observable. Also, the Smagorinsky model shows smaller fluctuations in the wake (chord 1), and the direct simulation shows larger fluctuations in the other three locations (chords 2, 3 and 4). This has an effect on the vibrations of the e-block arm, as will be discussed subsequently.

3.3.4   Energy Spectra

The spectrum of kinetic energy is useful in demonstrating the distribution of energy among the various scales of motion. For flows with simple geometries and/or periodic domains, obtaining a kinetic energy spectrum is straightforward. However, in our test case, the turbulence is inhomogeneous, and the mean flow is hard to define. In general, the problem does not lend itself to theoretical analysis. To obtain a turbulence spectrum we measure the azimuthal velocity at a particular point in the domain as a function of time. Using Taylor's frozen field hypothesis [1]

---

[1] Taylor's hypothesis is based on the assumption that the time scale of turbulent evolution is much slower than the time scale of the mean flow. This is valid if the fluctuations are comparably smaller than the mean flow. We can then assume that the turbulent field is "frozen" and is simply advected by the mean flow. In this analysis, we ensure that the standard deviation of the velocity is not more than 10% of the mean. The error in the kinetic energy spectrum associated with such an approximation is not easy to quantify.

We convert this time history to a spatial history, and use this data to obtain a (one dimensional, scalar) spatial auto-correlation function,

$$R(x) = \langle u_\phi(x_0) u_\phi(x_0 + x) \rangle \qquad (3.9)$$

where the brackets indicate averaging over all x_0. The Fourier transform of this function represents the one dimensional kinetic energy spectrum as a function of wavenumber (k). This turbulence spectrum of the airflow in a disk drive provides valuable confirmation of the existence of an inertial cascade.

According to Kolmogorov's law of universal equilibrium the energy spectrum E(k) should scale as,

$$E(k) = C_k \epsilon^{2/3} k^{-5/3} \qquad (3.10)$$

where C_k is a constant of order unity, ϵ is the dissipation rate (rate of energy transfer through the cascade process) and k is the wavenumber. In our case the dissipation ϵ (which traditionally has units of [length^2/time^3]) may be taken to be the windage per unit mass. It is hypothesized to be independent of the wavenumber k, hence we use the average value of windage for estimating it. These values of are in good agreement with our preliminary $k - \epsilon$ calculation. The values calculated by different SGS models are listed in Table 3.2.

Figure 3.26 shows the kinetic energy spectra obtained using the different models. These have been constructed by using velocity data at a single point in the wake of the arm. Spectra based velocity data at other points in the drive do not show significant differences from those in Figure 3.26. Firstly, all spectra show rough agreement with the $-5/3^{rd}$ law (see the thick line in Figure 3.26) demonstrating the existence of an inertial sub-range. We observe that the Smagorinsky model curve drops off faster than those of the dynamic model and the localized dynamic model indicating the dissipation of energy at higher wavenumbers. Also, the energy spectrum corresponding to the direct simulation contains the most energy at high wavenumbers, indicating an excessive build up of small scales due to the lack of an SGS model.

Theoretically, a more logical comparison can be made between LES and DNS energy spectra. If the filter function is known in wavenumber space an LES spectrum may be divided by the square of this function, to obtain the equivalent "unfiltered" spectrum. In our simulations, however, the top-hat filter is anisotropic and inhomogeneous in all three directions. The use of a one dimensional function to represent such a filter is not accurate and hence we refrain from making such a comparison.

There is also very little difference in the spectra predicted by the localized dynamic model and the dynamic model. The localized dynamic model has the advantage of computing a transport equation for SGS-k, which should include non-local and history effects. However, on a sufficiently fine grid such as ours the assumption of equality between production and dissipation appears to be valid, and very little difference is observed in the flow fields of the dynamic model and the localized dynamic model. Finally, we observe a significant variation in the model coefficients both in space and time, for the dynamic model (C) and the localized dynamic model (c_ν and c_ϵ). The Smagorinsky model is unable to capture this local variation. However, this spatial and temporal variation cannot be interpreted easily; hence we refrain from plotting it.

3.3.5   Vibrations

Often the off-track vibrations of the e-block arm or the slider are the most desired results of such a coupled fluid-structure simulation. Hence we compare the vibrations predicted by the various simulations. The off track vibrations of the tip of the e-block arm are plotted in Figure 3.27 as a function of time. The mean and peak-to-peak amplitudes of vibrations are given in Table 3.3. The mean is calculated by averaging over the final 6 revolutions of the disk, and the peak-to-peak is defined as the difference between the maximum and minimum deflection during this period. We observe that the Smagorinsky model, which predicts a slightly higher mean displacement also predicts the least peak to peak oscillations. Clearly, this is a direct result of the reduced pressure fluctuations. On the other hand, the direct simulation, due to its excessive fluctuations, records a smaller mean displacement and larger peak-to-peak oscillations. Although the differences in vibration values predicted by the simulations are small (less than 1 nm), we note that these trends will get amplified several times when more realistic structures such as suspensions and sliders are included in the simulation and the sliders off-track vibrations are compared.

Figure 3.28 shows the frequency spectra of the off-track vibrations shown in Figure 3.27. In all 4 cases we see the same modes (which correspond to sway and torsion) are excited in the structure. (Peaks are observed at 6.6 kHz, 7.5 kHz, 10 kHz, 1.12 kHz and 1.195 kHz.)

3.3.6   Comparison of computational cost

Large eddy simulations of disk drive airflows need to be computed until the turbulence field achieves a statistically steady state and sufficient time has elapsed for the important modes of the structure to be excited. This typically requires that the computations be carried out for 6-10 revolutions of the disk. Additionally the dynamics of interest lies in the 0-25 kHz range, which limits the size of the time step. As a result, such calculations take a substantial amount of time on desktop workstations, ranging from a couple of weeks to more than a month. In this context the cost of each turbulence model becomes important. In Table 3.4 we compare the normalized cost per time step of each turbulence model with a Navier Stokes solution on the same grid. This data has been obtained on a desktop Pentium 4 computer running at 3.2 GHz with 2 GB of RAM. The dynamic model is 33% more expensive than the Smagorinsky model and the localized dynamic model is 25% more expensive than the dynamic model.

3.4   Conclusions

The study of large eddy simulation SGS models is of considerable interest to the future research in airflow simulations in disk drives. This Chapter presents an investigation of three SGS models, under the limitations of a commercial CFD code. These models occur almost invariably in popular CFD software and their inclusion in a calculation is very easy. We provided *a posteriori* tests of the Smagorinsky model, the dynamic model and the localized dynamic model. By examining various turbulence statistics and measures like the kinetic energy and the energy spectrum we are able to draw useful conclusions about the performance of each model.

We conclude that the Smagorinsky model does not correlate well with the direct simulations in terms of mean and fluctuating velocities and pressures. We see a better correlation between the dynamic model, the localized dynamic model and the direct simulation. We also observe very little difference between the results predicted by the dynamic model and the localized dynamic model. The Smagorinsky model has a tendency to predict the highest dissipation at small scales, and this leads to smaller fluctuations in velocity and pressure. This extra dissipation leads to smaller peak-to-peak oscillations of the e-block arm, and we anticipate that the errors in vibration results of the structure would be amplified by the addition of slender and more flexible structures like the suspension and the slider. The direct simulation does not resolve up to the Kolmogorov scale, and hence it lacks a mechanism to dissipate energy, which would have ideally taken place at the Kolmogorov scale. This leads to excessive energy at small scales and results in larger fluctuations. Due to this unphysical feature the structure displays more peak-to-peak oscillations.

The best choice for turbulence modeling appears to be either the dynamic model or the localized dynamic model since they agree well with the direct simulation in the mean and do not show the over dissipation of the Smagorinsky model at the small scales. However, the localized dynamic model requires the computation of SGS-k, which makes it the most expensive choice. This cost is not justified when compared to the results of the dynamic model, and hence we advocate the use of the dynamic model in the following Chapters. As mentioned earlier, we cannot say which model delivers the "true" physical behavior, but our effort to compare the models has revealed significant differences between them.

This Chapter has exclusively used the CFD-ACE code for comparing SGS models. In the next Chapter, we broaden our investigation by comparing three different commercial codes and their implementation of various SGS models.

3.5   Tables

Table 3.1: Legend for figures

SGS Model	Line Type in figures
Smagorinsky model	Full line
Dynamic model	Dashed line
Localized dynamic model	Dotted line
Direct simulation	Dash-dotted line

Table 3.2: Average dissipation predicted by different SGS models

SGS Model	ϵ
Smagorinsky model	101212.701
Dynamic model	104219.471
Localized dynamic model	104448.123
Direct simulation	88461.554

.

Table 3.3: Mean and peak-to-peak vibrations of e-block arm tip as predicted by different SGS models

SGS Model	Mean (nm)	Peak-to-Peak (nm)
Smagorinsky model	3.3838	1.2138
Dynamic model	3.1425	1.4965
Localized dynamic model	3.2445	1.5947
Direct simulation	2.9156	1.7805

Table 3.4: Normalized cost of different SGS models per time step

Method	Normalized cost per time step
Direct simulation (same grid)	1
Smagorinsky model	1.253
Dynamic model	1.677
Localized dynamic model	2.1

3.6   Figures

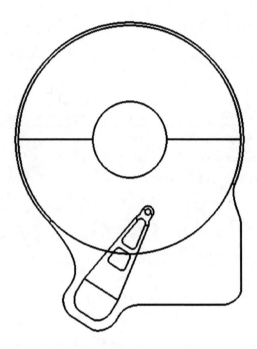

Figure 3.1: Top view of computational model

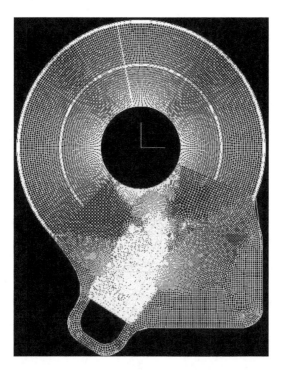

Figure 3.2: Plan view of hexahedral grid. The grid is block-structured in the symmetry region and grid density is increased upstream and downstream of the e-block arm

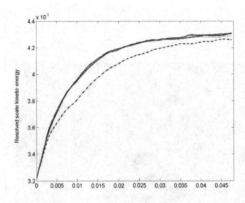

Figure 3.3: Resolved scale kinetic energy (see Table 3.1 for legend)

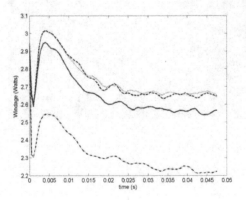

Figure 3.4: Windage (Watts) (see Table 3.1 for legend)

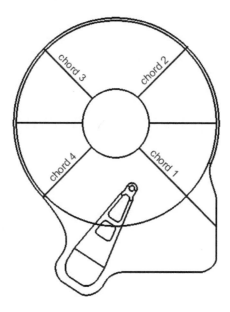

Figure 3.5: Location of radial lines (chords) for plotting turbulence intensity

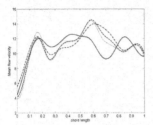

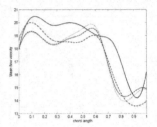

Figure 3.6: Mean azimuthal flow velocity along chord 1 (m/s) (see Table 3.1
for legend)

Figure 3.7: Mean azimuthal flow velocity along chord 2 (m/s) (see Table 3.1
for legend)

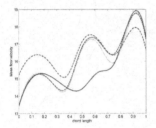

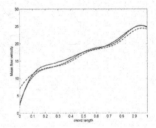

Figure 3.8: Mean azimuthal flow velocity along chord 3 (m/s) (see Table 3.1
for legend)

Figure 3.9: Mean azimuthal flow velocity along chord 4 (m/s) (see Table 3.1
for legend)

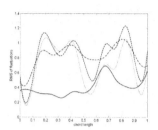

Figure 3.10: RMS fluctuating azimuthal flow velocity along chord 1 (m/s) (see Table 3.1 for legend)

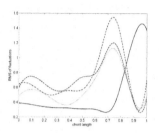

Figure 3.11: RMS fluctuating azimuthal flow velocity along chord 2 (m/s) (see Table 3.1 for legend)

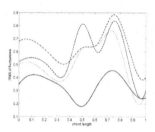

Figure 3.12: RMS fluctuating azimuthal flow velocity along chord 3 (m/s) (see Table 3.1 for legend)

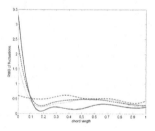

Figure 3.13: RMS fluctuating azimuthal flow velocity along chord 4 (m/s) (see Table 3.1 for legend)

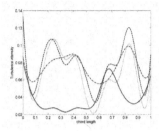

Figure 3.14: Turbulence Intensity along chord 1 (see Table 3.1 for legend)

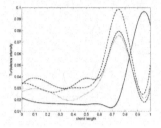

Figure 3.15: Turbulence Intensity along chord 2 (see Table 3.1 for legend)

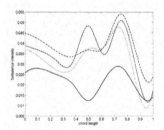

Figure 3.16: Turbulence Intensity along chord 3 (see Table 3.1 for legend)

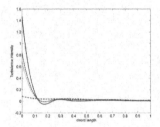

Figure 3.17: Turbulence Intensity along chord 4 (see Table 3.1 for legend)

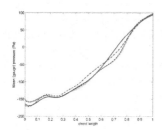

Figure 3.18: Mean pressure along chord 1 (Pa) (see Table 3.1 for legend)

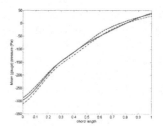

Figure 3.19: Mean pressure along chord 2 (Pa) (see Table 3.1 for legend)

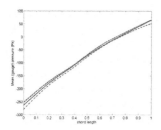

Figure 3.20: Mean pressure along chord 3 (Pa) (see Table 3.1 for legend)

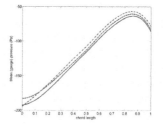

Figure 3.21: Mean pressure along chord 4 (Pa) (see Table 3.1 for legend)

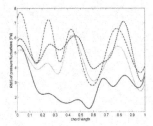

Figure 3.22: RMS of pressure fluctuation along chord 1 (Pa) (see Table 3.1 for legend)

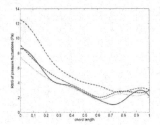

Figure 3.23: RMS of pressure fluctuation along chord 2 (Pa) (see Table 3.1 for legend)

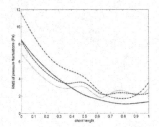

Figure 3.24: RMS of pressure fluctuation along chord 3 (Pa) (see Table 3.1 for legend)

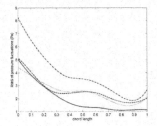

Figure 3.25: RMS of pressure fluctuation along chord 4 (Pa) (see Table 3.1 for legend)

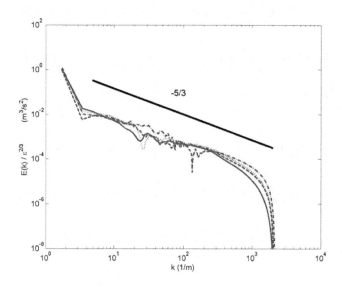

Figure 3.26: Kinetic energy spectra

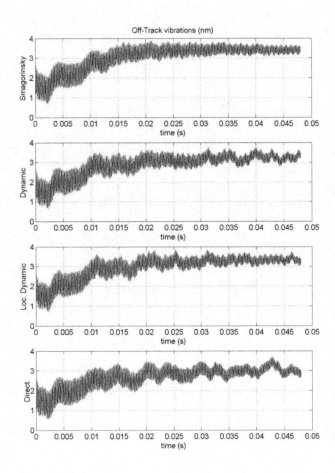

Figure 3.27: Off-Track vibrations of e-block arm tip

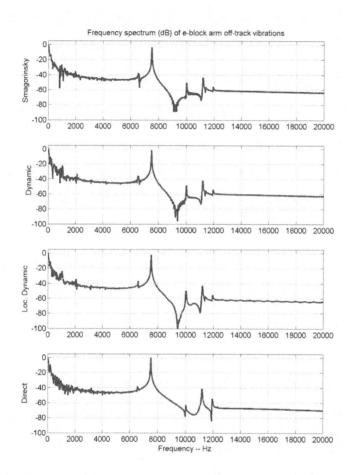

Figure 3.28: Frequency spectra of e-block arm off-track vibration

Chapter 4

A Comparison of Commercial CFD
Software

4.1 Introduction

Engineering calculations of flows in complex geometries such as ours are often presented without any verification or validation, in spite of being most susceptible to errors. (Validation is defined (Stern et al. 2001) as *"a process for ascertaining simulation modeling uncertainty by using benchmark experimental data and, when conditions permit, estimating the sign and magnitude of the modeling error itself")*. Moreover, commercial CFD codes used in industrial applications are efficient in calculating results, but do not offer insights into the numerical uncertainties of those results. The industrial CFD community usually treat commercial CFD codes as *"black boxes"* which return results given a set of inputs. In this light the LES results of the flow across a square prism are presented using three commercial codes, and four different SGS models.

From Chapters 2 and 3 we realize that flows is hard disk drives are highly complex and little experimental data is available for comparison (See Chapters 5 and 6 for experimental validation). When experimental data is limited a common practice is to use the commercial code to solve a well known test case for which a rich set of experimental data is available. This often helps in highlighting the merits and demerits of the software. The test case chosen here (the flow across a square prism) has several similarities with the original disk drive problem of interest. Both flows have a blunt body obstruction, massive flow separation, formation of a "flapping" shear layer, regions of laminar, transitional and turbulent flow, recirculation, vortex shedding and an inherent three dimensional nature.

In a CFD Biathlon Forum (Freitas 1995) several commercial codes were tested (including CFD-ACE and Fluent) in solving five model problems. The flow across a square prism was a part of these tests, but only 2-dimensional simulations using RANS models were presented. Several LES codes were used to solve the square prism case during a workshop held in Germany in June 1995 (Rodi et al. 1997). Similarly, at the ERCOFTAC Workshop, seven groups submitted their time-averaged solutions of the problem; the results are published in Voke (1997). In addition to these workshops there is a wealth of literature discussing the application of LES to this problem. Among the most recent works is that of Sohankar et al. (2000), who tested three SGS models and varied other parameters such as the grid size, time step and spanwise dimension. Fureby et al. (2000) also tested several SGS models and generated a database of first and second order statistical moments of the resolved velocity.

Most of the cited literature above use codes that were developed by university researchers, but there exist no such published tests on commercial codes. The main objective of this Chapter is to investigate the ability of three commercial codes to solve the square prism problem: CFD-ACE 2004, Fluent 6.2.16 and CFX 5.7.1,(and their implementations of four SGS models: the Smagorinsky model (Smagorinsky 1963), the (Algebraic) dynamic model (Germano et al. 1991), the localized dynamic one-equation model (Kim and Menon 1995) and the wall-adapting local eddy-viscosity (WALE) model (Nicoud and Ducros 1999)). Such a comparison between the simulation results from different commercial codes serves as an effective test of the internal numerics of the code, which are usually hidden from the user, e.g.

  – Segregated v/s Coupled solution strategies

  – Convergence criteria for each time step

  – Under-relaxation parameters in iterative segregated solvers

  – Artificially imposed bounds or limiters that prevent the code from crashing

  – Inaccurate representation of SGS models

4.2   Configuration and Setup

A two-dimensional schematic plan view drawing of the problem geometry is shown in Figure 4.1. In Cartesian coordinates, the origin is located at the center of the prism, the mean flow is oriented in the x-direction, and Figure 4.1 depicts a representative $x - z$ plane. The length of the square prism side is (in the streamwise direction) d and the inflow x-velocity is U_∞. Henceforth, as is traditional, all dimensions are scaled by d, all velocities by U_∞ and times by d/U_∞.

The Reynolds number of the flow ($Re = \frac{U_\infty d}{\nu}$) was 22,000, the upstream distance, X_u was 4.5, while the downstream distance X_d was 15. The lateral dimension H was 4, while the dimension in the y-direction was 14. All of the simulations used the same grid, consisting of $165 \times 105 \times 25$ cells, an $x - y$ plane of which is shown in Figure 4.2. The distribution of nodes was uniform outside a region extending two units upstream, downstream and sideways (in the y-direction) of the prism (as in Sohankar et al. (2000)). The uniform cell spacing was

0.167 downstream (Δx_d), 0.25 upstream (Δx_u) and 0.167 in the z-direction (Δz), again as in Sohankar et al. (2000). In the region of the grid close to the prism, a hyperbolic tangent function was used to stretch the cells. The first node away from the prism wall was at a distance of 0.00815 in both and x- and y- directions.

The CFD-ACE and Fluent codes are based on the incompressible cell-centered finite volume formulation. The governing system was solved iteratively using the SIMPLEC technique (originally due to Van doormaal and Raithby 1984) (i.e. they use segregated solvers), although other methods (e.g. PISO) were available in Fluent. On the other hand, CFX employs a pressure based coupled solver. A preconditioned multigrid method is used to solve the linear system arising from the coupled Navier Stokes and continuity equations. For further details on the solution strategies of each software, we refer the reader to their respective user manuals. Seven simulations were computed:

1. CFD-ACE using the Smagorinsky model (ACE1)

2. CFD-ACE using the dynamic model (ACE2)

3. CFD-ACE using the localized dynamic model (ACE3)

4. Fluent using the Smagorinsky model (Flu1)

5. Fluent using the dynamic model (Flu2)

6. Fluent using the WALE model (Flu3), and,

7. CFX using the Smagorinsky model (CFX1) [1]

Descriptions of the above SGS models are given in Sections 2.5.1 to 2.5.3. Simulations ACE1, Flu1 and CFX1 used $C_s = 0.1$ for the Smagorinsky's model. In simulation Flu2, the value of C_s was clipped below 0 and above 0.23. And in simulation Flu3, the parameter C_ω of the WALE model was set to 0.325.

All simulations used centered differencing for the convective terms to avoid the well known diffusion associated with upwind biased schemes. The effect of adding a small amount of upwind-biased differencing is discussed later in Section 4.4. All CFD-ACE simulations used the first order Implicit Euler's method for time advancement. A semi-implicit second order method (Crank Nicholson) was available in CFD-ACE, but calculations using it became unstable as time progressed. All Fluent simulations and the CFX simulation used a two step BDF method, which is second order accurate and provides better stiff stability than corresponding implicit Adams methods (This method is also known as BDF2 or Second Order Implicit Euler). For both Fluent and CFX it not not clear from their user manuals how these 2 step methods are started.

The inflow boundary condition was specified to be a constant inflow in the x-direction ($u = U_\infty$, $v = 0$, $w = 0$), perturbed with 2% turbulent fluctuations. In Fluent the

---

[1] The Smagorinsky model was the only SGS model available in CFX.

"spectral synthesizer" model (based on Smirnov et al. 2001) was used to model the velocity fluctuations. At the outflow convective boundary conditions of the form

$$\frac{\partial u_i}{\partial t} + U_\infty \frac{\partial u_i}{\partial x} = 0 \qquad i = 1, 2, 3, \tag{4.1}$$

were used. Symmetry conditions simulating a frictionless wall,

$$u_n = \left(\frac{\partial u_i}{\partial n}\right)_{i \neq n} = 0, \tag{4.2}$$

were used for all of the exterior lateral walls, where n is the normal direction to the wall.

The prism wall was modeled as a no-slip boundary in all simulations. None of the simulations used wall functions and the coarseness of the grid did not allow the very small turbulent structures near the wall to be resolved. Simulations ACE1 and CFX1 used the standard Van Driest (1956) damping modification for the Smagorinsky parameter C_s near the wall. Flu1 and Flu2 used a "damped mixing length" near the wall, such that for the Smagorinsky model constant, $C_s = \min(\kappa y_{\text{wall}}, 0.1\Delta)$, where κ is the von Kármán constant and Δ is the filter width. The other simulations, ACE2, ACE3 and Flu3 did not use any near wall modeling. All simulations used implicit grid filtering for the Smagorinsky models, and used a top-hat filter (which is anisotropic and inhomogeneous) for test filtering in the dynamic models, whose size was twice the grid filter.

4.3   Results and Discussion

All simulations were started from the initial conditions of rest and ran for at least 8 shedding cycles, identified by the time history of the lift. Coherent vortex shedding started after approximately 500 time steps. Flu1, Flu2, Flu3 and CFX1 were computed on local desktop Pentium machines, while ACE1, ACE2 and ACE3 were computed on a parallel cluster using 8 processors.

Two sets of results are presented in this Chapter: time-averaged and phase averaged. Phase averaged data was not available for CFX1, hence only time-averaged data will be presented for it. Time averaging was done only over complete shedding cycles i.e. initial start up data was ignored. Phase averaged data is presented by breaking up each shedding cycle into twenty phases, as in Lyn and Rodi (1994) and Lyn et al. (1995).

4.3.1   Global Quantities

Table 4.1 reports global quantities of the flow. $St = \frac{fd}{U_\infty}$ is the Strouhal number, l_r is the time-averaged recirculation length (calculated from the prism center), $\langle C_D \rangle, C_{D,\text{RMS}}, \langle C_L \rangle, C_{L,\text{RMS}}$ are the mean and RMS. values of the coefficients of drag and lift on the prism, respectively. Some authors (of both numerical and experimental works) choose to report global quantities corrected for blockage effects ((Sohankar et al. 2000), Bearman and Obasaju (1982)).

However, both LES workshops (Rodi et al. (1997) and Voke (1997)) do not present blockage-corrected results, and this custom is followed here also. It should be noted that obtaining blockage-corrected values from the results presented here is a straightforward exercise, given that the blockage parameter is 7.1% (the ratio of the projected area of the prism to the area of the empty channel).

Table 4.1 includes results from our test cases (ACE1 to CFX1) and results from several experimental investigations. Although some of the experiments used vastly different Reynold's numbers a rough comparison still holds, based on the grounds that non-dimensional quantities like force coefficients are independent of the Reynolds number once the Reynolds number is above 20,000 (McLean and Gartshore 1992). Also included are three representative results from the LES workshop in Germany (Rodi et al. 1997) and direct numerical simulation (DNS) result from the workshop by Voke (1997).

It appears that all of our simulations are accurate in predicting the Strouhal number while not being accurate in other quantities, which confirms the idea that the Strouhal number is insensitive to the SGS model. The mean recirculation length, which is an important quantity that determines the average size of the wake, is computed from the time-averaged streamwise velocity profiles. As will be evident from the velocity profile itself, Flu2 and ACE2 most accurately predict l_r. Flu2 is the better of the two predictions, while the worst result is from CFX1. All simulations overpredict the mean drag coefficient when compared to the experiments, but the dynamic models (ACE2, ACE3 and Flu2), which do not use any near-wall damping, are better at predicting the mean drag than the Smagorinsky models. Generally it is expected that the recirculation length and the mean drag coefficient are inversely proportional, but the mean drag values are close to each other, and a clear trend is not manifested. It is also known (due to Lee 1975) that increasing the free stream turbulence decreases the mean drag. Even though all the experimental results are for free streams which are *smooth* and relatively less turbulent (except Lyn and Rodi (1994) which report 2% upstream turbulence), our simulations predict a higher mean drag. With the exception of ACE2, ACE3 and CFX1 there is good agreement in the RMS drag coefficient, while the mean lift coefficient (which should be zero due to symmetry) is appropriately close to zero in all simulations. For flow structure interaction problems it is crucial to predict the RMS lift coefficient accurately. Generally the RMS lift coefficient is determined by the vortex dynamics of the wake since the lift is directly related to changes in circulation around the prism. The dynamic models, ACE2 and Flu2, again appear to provide impressive results, with ACE2 the better of the two simulations. Both the WALE model and the localized dynamic model under predict $C_{L,RMS}$, but there is no consistent trend among the Smagorinsky models: ACE1 overpredicts $C_{L,RMS}$, while Flu1 and CFX1 underpredict this quantity.

In conclusion, ACE2 and Flu2, both based on the dynamic model, appear to provide the best agreement regarding the important global quantities. The two other models tested here, in simulations ACE3 and Flu3, provide reasonable agreement in all global quantities, but they under predict the RMS lift coefficient.

In the remainder of the Chapter a detailed comparison is provided with the results of Lyn and Rodi (1994). However, as pointed out in Sohankar et al. (2000), such a comparison should be made with caution. This is because the experimental measurements were made without the use of "end plates" and the prism aspect ratio used in the experiments was relatively small (side = 1 : axial length = 9.75).

4.3.2   Time-Averaged Quantities

The time-averaged streamwise velocity along the centerline is plotted in Figure 4.3. The legend is given in Table 4.2, and it is used in all subsequent figures. The mean recirculation length, which is the point of zero-crossing of the streamwise velocity, has already been discussed. In the near wake region all simulations tend to overpredict the size of the wake. Of the ACE simulations, ACE2, based on the dynamic model has the best spatial agreement with the experiments. Among the Fluent simulations, Flu2, again based on the dynamic model, has excellent agreement with the experimental data, better than all other simulations. Simulations using the Smagorinsky model (CFX1, ACE1 and Flu1) consistently overpredict the negative velocity in the wake. The experimental data shows that the velocity levels off quickly at about 4 span lengths to about 60% of the free stream velocity. This trend is not displayed by any of the simulations; all simulations tend to level off at much later distances, to larger values. This has been a common trend in much of the published simulations (at least Sohankar et al. (2000), Rodi et al. (1997) and Voke (1997)). The reasoning behind such a trend is unclear: the SGS model, grid stretching and freestream turbulence may play a part.

Figures 4.4 and 4.5 show the variation of the RMS velocities with the streamwise length. These velocities are thus time-averaged representations of the Reynolds stress tensor. Since LES does not explicitly represent the small scales, but only represents their effect on the large scales through an SGS model, one cannot expect true agreement between the LES data and the experimental data. In general the agreement should increase if the higher frequency contribution to the RMS is negligible. Almost all simulations tend to overpredict the RMS streamwise velocity and underpredict the RMS vertical velocity. This trend (consistent among all simulations) indicates that the larger eddies of the flow, which are explicitly represented, show artificially higher fluctuations in the direction of the mean flow and smaller fluctuations in secondary directions orthogonal to the mean flow. The spatial distribution of the RMS streamwise velocity (e.g. the location of the peak) is also likely to be influenced by the mean flow. The Smagorinsky solutions of ACE (ACE1) and Fluent (Flu1) show the highest RMS velocities in both the streamwise and vertical directions. A correlation is clearly evident between the RMS streamwise velocity in the wake and the RMS lift on the prism.

The cross term of the time-averaged Reynolds stress $\langle u'v' \rangle$, which is a measure of the anisotropy of the turbulent field, is shown in Figure 4.6. Among the various quantities discussed in this Chapter the cross term of the Reynolds stress is generally the most difficult for any SGS model to accurately reproduce. ACE3 shows excellent agreement with the experimental data, and Flu2 and Flu3 also show good agreement. The central advantage of the localized dynamic model (of ACE3) over the algebraic dynamic models (of ACE2, Flu2) is that it captures the "non-local and history effects" of the flow by computing the differential equation for the SGS kinetic energy. From Figure 4.6 it appears that this model has a superior ability to predict the cross term of the Reynolds stress, hence providing a better representation of the anisotropy of the flow. ACE1, Flu1 and CFX1 (all using the Smagorinsky model) show the poorest agreement with the experimental data. In addition to the magnitude of the cross term of the Reynolds stress, the sign of this term is also important. The sign of this term (along with the velocity gradient of the mean flow) determines the production or loss of turbulent kinetic energy (sometimes referred to as "shear production"). It is important to correctly represent the interaction between the mean flow and the turbulent field, and ACE2, ACE3, Flu2 and Flu3 are superior to the the Smagorinsky models in this regard.

4.3.3 Phase Averaged Quantities

In the original work of Lyn and Rodi (1994) phase definitions were based on the peaks in the pressure signal obtained from a piezoelectric pressure transducer at the center of the prism sidewall. In our simulations, since the wall region is not computed completely, we choose not to rely on the peaks in the pressure at one point on the prism side wall. On the other hand, peaks in the global lift spectrum, which is an integral of the pressure on all the prism walls, do not directly correspond to a peak in the pressure signal of Lyn and Rodi (1994). Due to this difficulty in demarcating phases the vertical velocity was used as an indicator for phase definition. Each shedding cycle was separated into 20 phase bins and ensemble averaging was performed. Phase 01 was then assigned to the bin with the most agreement (with Lyn and Rodi (1994)) in the vertical velocity and all other phases were numbered successively. In all cases, Phase 01 turned out to be one phase bin beyond the negative peak in the lift time history. This is roughly consistent with Lyn and Rodi (1994) since a peak in pressure on the top face of the prism corresponds roughly to a negative peak in the lift history. Finally, the original idea, that the first half cycle corresponds to an accelerating free stream (adjacent to the top side wall) and the second half corresponds to a decelerating free stream, still holds in our simulation phases.

Figures 4.7- 4.12 show the phase averaged streamlines of the flow, depicting Phase 01. Since the numerous vortices in the near wall region are not captured in the calculations, and the streamlines are created from interpolated velocity values on a coarse grid, the region close to the prism walls should be ignored. For reference, corresponding streamline pictures are also shown for the experimental results of Lyn and Rodi (1994) in Figure 4.13 and the RANS calculations of Lakehal and Thiele (2001) are shown in Figure 4.14 [2]. In general there is very good qualitative agreement of the simulations with the experiments. Similar figures for Phase 09 are depicted in Figures 4.15- 4.20. The experimental results of Lyn and Rodi (1994) are shown in Figure 4.21, and the RANS calculations of Lakehal and Thiele (2001) are in Figure 4.22. The phase sorted data presented here helps in understanding several features of the flow that cannot be deduced from the time-averaged data only.

One of the attributes of interest in the streamlines for Phase 01 is the location of the

---

[2] Although this calculation is not an LES, it is among the few published streamline pictures, and hence is reproduced here

streamline on the top of the prism that separates the shed vortex from the free stream. This streamline depicts the amount of vertical oscillation of the wake, and a consistent connection is evident with the RMS of the lift. Larger oscillations of the wake, as in ACE1 (Figure 4.7), lead to larger lift coefficients, while smaller oscillations, as in ACE3 and Flu3 (Figures 4.9 and 4.12), lead to smaller lift coefficients.

Another attribute of interest for Phase 01 is the location of the same streamline below the prism that does not get entrained in the wake. Again, a correlation is observable between the location of this streamline and the mean drag on the prism. Cases in which this streamline is closer to the back face of the prism (thus predicting a smaller shed vortex during Phase 01) correspond to cases with higher mean drag forces (ACE1), while the reverse is also true (ACE3)

For Phase 09 it appears that the separating streamline below the prism that is not entrained in the wake is located too far below the prism in ACE1 (Figures 4.15) but too close to the prism in ACE3 (Figures 4.9). This correlates well with the corresponding lift coefficients. In general ACE2, Flu2 and Flu3 show the best agreement with the experimental data.

Figures 4.23 and 4.24 show the variation of the vertical velocity along the centerline, for phases 01 and 09, respectively. Agreement of the vertical velocity for Phase 01 was used as a method to sort phases. In the near wake ACE1 clearly predicts more severe values of vertical velocity (both positive and negative), while ACE3 and Flu3 show much smaller values. This is consistent with the over- and under- estimation of the oscillations in the wake for ACE1, and ACE3, Flu3 respectively. ACE2 and Flu2 show excellent agreement for Phase 01, but by Phase 09 the agreement of Flu2 is much reduced.

In addition to Phases 01 and 09 similar figures are shown for two intermediate phases: phase 05 and 15 (Figures 4.25 and 4.26). Phases 05 and 15 are among the "accelerating" and "decelerating" phases, respectively, since the free stream adjacent to the top prism side walls accelerates during Phase 05 and decelerates during Phase 15. During these phases also simulations using the Smagorinsky models overpredict the positive and negative vertical velocities. For these phases the agreement of ACE2 with the experimental data is remarkably good, while none of the other simulations come within close agreement of the experiment.

4.4 The Effect of Upwind differencing

It is often claimed that first order upwind differencing (applied to the convective term in the standard finite volume formulation) produces artificial dissipation which makes it unsuitable for large eddy simulation (Mittal and Moin 1997). Two simulations from our study above (ACE1 and ACE2) were recomputed with the addition of 10% upwinding to the differencing scheme of the convective terms. In these simulations the final difference is the sum of 90% contribution from central differencing and 10% contribution from upwind differencing. It should be noted that this technique of "blending" the original difference with upwinding is a default setting in the CFD-ACE code.

Table 4.3 shows the change in the global quantities of the flow due to the introduction of upwinding. Figures 4.27 and 4.28 show the change in the time averaged x-velocity for ACE1 and ACE2, respectively. Figures 4.29, 4.30 and 4.31, 4.32 show the change in RMS streamwise and vertical velocity, respectively. For completeness, the streamlines of the flow for Phase 01 and Phase 09 are shown in Figures 4.33–4.36 and Figures 4.37–4.40, respectively.

In both cases, on adding upwinding, the Strouhal number is slightly decreased and the RMS lift coefficient is increased (due to larger oscillations of the wake). Another common observation is that the initial time required for the start of vortex shedding is increased.

For the simulation ACE1 the length of the recirculation zone is almost unchanged as is also evident from the streamwise velocity profile in Figure 4.27. The slight increase in l_r is associated with a slight decrease in the mean drag $\bar{C}_D$. The small increase also occurs in the RMS coefficient of the lift, but the change in the RMS velocity fluctuations is negligible. Additionally, the streamline pictures show that vertical deflection of the streamlines due to the formation of the vortex at Phase 01 or 09 is almost negligible.

For the simulation ACE2 the length of the recirculation zone is decreased significantly (see Figure 4.28), and correspondingly the mean drag coefficient increases. The RMS streamwise and vertical velocities show considerable increases with the addition of upwinding and this results in the higher RMS lift coefficient. Larger oscillations of the wake are also evident in the streamline pictures for Phase 01 and Phase 09.

One would expect that for a fixed given inlet kinetic energy the addition of numerical dissipation would reduce the actual energy of the flow, possibly leading to smaller fluctuations. However, it is difficult to interpret the above results on the basis of the reduced kinetic energy of the flow alone, since SGS and viscous dissipation also change when the spatial features of the flow change. Calculating the SGS and viscous dissipation is not an easy task in most commercial codes. While estimates of a particular source of dissipation (e.g. numerical) may be obtained by turning off the other sources (e.g. SGS and viscous), such estimates cannot be obtained for the entire length of the calculation. In conclusion, the effect of upwinding on the flow may be summarized as follows:

1. For the Smagorinsky model (ACE1), most features of the flow remain unchanged, while there was a small increase in the lift coefficient. A possible explanation for this observation is that the incremental dissipation introduced by the upwinding is very small compared to the other (SGS, viscous and numerical) forms of dissipation.

2. For the dynamic model (ACE2) there was a significant change in the flow features. The recirculation zone is shortened, thereby increasing the drag on the prism. The wake oscillates more vertically, leading to higher RMS lift coefficients.

4.5   Concluding Remarks

In this Chapter the flow across a square prism has been calculated using LES with three different commercial codes employing 4 SGS models. The results were benchmarked using the well known test case of Lyn and Rodi (1994). The effect of the addition of upwind differencing was also studied in two of the simulations. Although the study does not examine the flows in disk drives several important features of commercial codes that may be used in that application have been brought to light. The main conclusions drawn through this investigation are:

1. The Strouhal number is not an indicator of an accurate simulation, since an accurate Strouhal number does not translate to accuracy in other features of the flow.

2. The dynamic models (ACE2 and Flu2) provide impressive agreement in the recirculation length and the RMS of the lift coefficient, which are the two most important global quantities of the flow

3. The dynamic models (especially Flu2) again provide the best agreement in the time averaged streamwise velocity

4. All simulations tend to over-predict the streamwise velocity fluctuations and under-predict the vertical velocity fluctuations. Higher velocity fluctuations, especially using the Smagorinsky model, correlate well with higher lift coefficients

5. When the time dependant data is split into phase bins and ensemble averaged several features of the flow come to light, e.g. the vertical oscillation of the wake, the size and position of the shed vortex, etc. In general Flu2 and ACE2 offer the best spatial prediction of the wake during its different phases. Correlations can be readily made from the spatial structure of the wake during certain phases, and the global time-averaged results of lift and drag.

6. The addition of upwind differencing has marginal effects on the simulations using the Smagorinsky model but more dramatic effects on the simulations using the dynamic model. In both cases the shedding process is slowed slightly and the oscillation of the wake is increased, leading to artificially higher lift coefficients.

7. Finally, the overall performance of CFX's implementation of the Smagorinsky model is poor compared to CFD-ACE and Fluent. This is a direct indicator of the internal numerics of the code

In the next Chapter we revert back to disk drive flows and estimate the role of the grid in the accuracy of solutions. Chapters 5 and 6 also provide crucial validation against two experimental datasets.

4.6 Tables

Table 4.1: Global Results

Current Work	$Re/10^3$	St	l_r	$\langle C_D \rangle$	$C_{D,\mathrm{RMS}}$	$\langle C_L \rangle$	$C_{L,\mathrm{RMS}}$
ACE1	22	0.132	1.715	2.422	0.211	-0.09	1.578
ACE2	22	0.132	1.515	2.132	0.138	0.006	1.280
ACE3	22	0.130	1.626	2.044	0.123	0.001	1.056
Flu1	22	0.129	1.604	2.309	0.192	0.027	1.142
Flu2	22	0.130	1.404	2.210	0.213	-0.151	1.373
Flu3	22	0.130	1.554	2.260	0.259	-0.050	1.064
CFX1	22	0.130	2.627	1.931	0.125	-0.01	1.201
Experiments							
Lee (1975)	176	0.122	-	2.05	0.23	-	1.24
Vickery (1966)	100	0.12	-	2.05	0.17	-	1.32[3]
Lyn and Rodi (1994)	21.4	0.134	1.38	2.1[4]	-	-	-
Bearman and Obasaju (1982)	22	0.13	-	2.1	-	-	1.327[5]
Norberg (1993)	13	0.13	-	2.16	-	-	-
McLean and Gartshore (1992)	23	0.13	-	-	-	-	1.3
From Rodi et al. (1997)							
IIS-KOBA	22	0.13	1.22	2.04	0.26	-0.3	1.31
UKAHY1	22	0.13	1.32	2.2	0.14	-0.02	1.01
TAMU1	22	0.13	1.15	2.28	0.2	-0.03	1.37
From Voke (1997)							
DNS[6]	22	0.133	-	2.09	0.178	0.005	1.45

---

[3]For a *smooth* stream with low turbulent fluctuations

[4]The mean drag coefficient was estimated by integrating the momentum flux based on the mean velocity

[5]Original value reported was 1.2, after correcting for blockage

[6]Data based on three shedding cycles only

Table 4.2: Common Legend for Figures

Simulation	Marker
ACE1	.
ACE2	○
ACE3	×
Flu1	+
Flu2	*
Flu3	◇
CFX1	▽
Experiments of Lyn and Rodi (1994)	□
ACE1 with 10% upwinding	▷
ACE2 with 10% upwinding	◁

Table 4.3: The effect of 10% upwind differencing

Case	St	l_r	$\langle C_D \rangle$	$C_{D,\mathrm{RMS}}$	$\langle C_L \rangle$	$C_{L,\mathrm{RMS}}$
ACE1	0.132	1.715	2.422	0.211	-0.09	1.578
ACE1 with upwinding	0.128	1.778	2.391	0.215	-0.089	1.714
ACE2	0.132	1.515	2.132	0.138	0.006	1.280
ACE2 with upwinding	0.127	1.169	2.428	0.225	-0.003	1.711

4.7   Figures

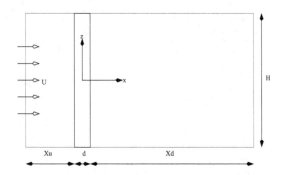

Figure 4.1: Model Configuration and Setup

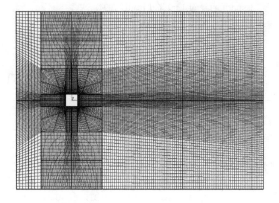

Figure 4.2: Cross Section of the Grid in the x-y plane. The grid is uniform in the axial (z)
direction

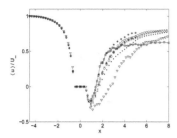

Figure 4.3: Time averaged streamwise velocity, non-dimensionalized by the free stream velocity U_∞. See Table 4.2 for legend

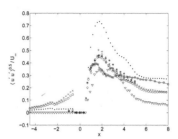

Figure 4.4: Time averaged RMS streamwise velocity, non-dimensionalized by the free stream velocity U_∞. This is also the square root of the $(1,1)$ normal Reynolds stress. See Table 4.2 for legend

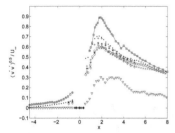

Figure 4.5: Time averaged RMS vertical velocity, non-dimensionalized by the free stream velocity U_∞. This is also the square root of the $(2,2)$ normal Reynolds stress. See Table 4.2 for legend

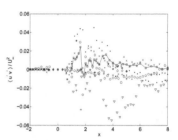

Figure 4.6: Time averaged cross term $(1,2)$ of the Reynolds stress tensor, $\langle u'v' \rangle$, non-dimensionalized by U_∞^2. See Table 4.2 for legend

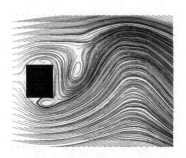

Figure 4.7: Streamlines for Phase 01, ACE1

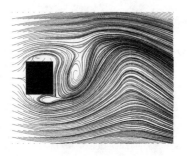

Figure 4.8: Streamlines for Phase 01, ACE2

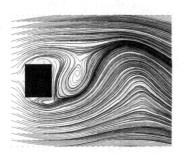

Figure 4.9: Streamlines for Phase 01, ACE3

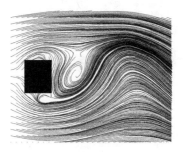

Figure 4.10: Streamlines for Phase 01, Flu1

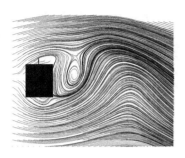

Figure 4.11: Streamlines for Phase 01,
Flu2

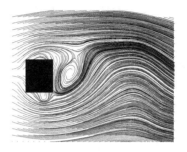

Figure 4.12: Streamlines for Phase 01,
Flu3

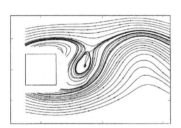

Figure 4.13: Streamlines for Phase 01,
Lyn and Rodi (1994)

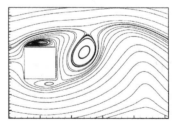

Figure 4.14: Streamlines for Phase 01,
from Lakehal and Thiele (2001)

Figure 4.15: Streamlines for Phase 09, ACE1

Figure 4.16: Streamlines for Phase 09, ACE2

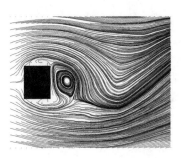

Figure 4.17: Streamlines for Phase 09, ACE3

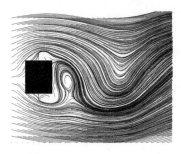

Figure 4.18: Streamlines for Phase 09, Flu1

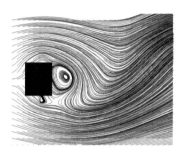

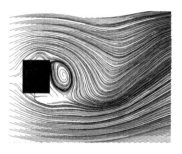

Figure 4.19: Streamlines for Phase 09, Flu2

Figure 4.20: Streamlines for Phase 09, Flu3

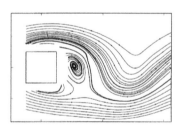

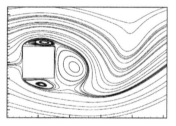

Figure 4.21: Streamlines for Phase 09, Lyn and Rodi (1994)

Figure 4.22: Streamlines for Phase 09, from Lakehal and Thiele (2001)

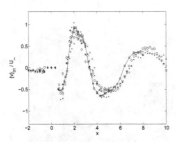

Figure 4.23: Phase averaged vertical velocity for Phase 01, non-
dimensionalized by U_∞. See Table 4.2
for legend

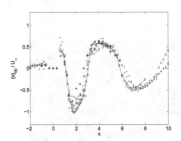

Figure 4.24: Phase averaged vertical velocity for Phase 09, non-
dimensionalized by U_∞. See Table 4.2
for legend

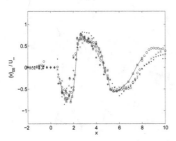

Figure 4.25: Phase averaged vertical velocity for Phase 05, non-
dimensionalized by U_∞. See Table 4.2
for legend

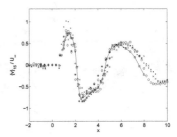

Figure 4.26: Phase averaged vertical velocity for Phase 15, non-
dimensionalized by U_∞. See Table 4.2
for legend

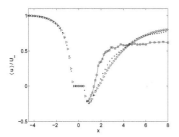

Figure 4.27: Comparison of time averaged streamwise velocity between ACE1 and ACE1 with 10% upwind differencing. See Table 4.2 for legend

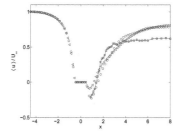

Figure 4.28: Comparison of time averaged streamwise velocity between ACE2 and ACE2 with 10% upwind differencing. See Table 4.2 for legend

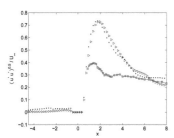

Figure 4.29: Comparison of time averaged RMS streamwise velocity between ACE1 and ACE1 with 10% upwind differencing. See Table 4.2 for legend

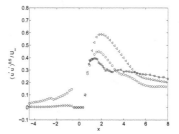

Figure 4.30: Comparison of time averaged RMS streamwise velocity between ACE2 and ACE2 with 10% upwind differencing. See Table 4.2 for legend

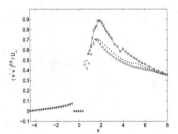

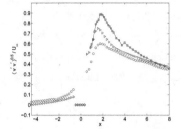

Figure 4.31: Comparison of time averaged RMS vertical velocity between
ACE1 and ACE1 with 10% upwind differencing. See Table 4.2 for legend

Figure 4.32: Comparison of time averaged RMS vertical velocity between
ACE2 and ACE2 with 10% upwind differencing. See Table 4.2 for legend

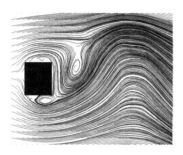

Figure 4.33: Streamlines for Phase 01, ACE1

Figure 4.34: Streamlines for Phase 01, ACE1 with 10% upwinding

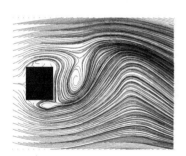

Figure 4.35: Streamlines for Phase 01, ACE2

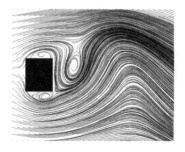

Figure 4.36: Streamlines for Phase 01, ACE2 with 10% upwinding

Figure 4.37: Streamlines for Phase 09, ACE1

Figure 4.38: Streamlines for Phase 09, ACE1 with 10% upwinding

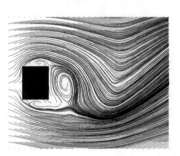

Figure 4.39: Streamlines for Phase 09, ACE2

Figure 4.40: Streamlines for Phase 09, ACE2 with 10% upwinding

Chapter 5

Grid Convergence & Experimental Validation

5.1 Introduction

As mentioned in Chapter 4 in most numerical studies of complex flows such as those in hard disk drives little attention is paid to reporting the numerical errors and uncertainties of the results. While it has become relatively easy to calculate such flows using commercial CFD software, the accuracy of these results is questionable at best. In the current Chapter we hope to shed some light on the sources of discrepancies between numerical and experimental results on disk drive flows.

Errors (i.e. the difference between the simulation result and the actual physical value) may be divided into two broad parts: modeling errors and numerical errors. Modeling errors are due to mathematical assumptions of the physical problem itself; e.g. the assumption of incompressibility, the application of simplified boundary conditions, the use of a sub-grid scale turbulence model, the assumption of isothermal flow, etc. Numerical errors are those due to the technique of solving the mathematical problem; e.g. discretization in space and time, grid convergence, artificial dissipation and dispersion, truncation of the iteration process in every time step, computer round off, etc. By providing an estimate for each of these sources of errors simulation results may be corrected, and accurate results may be reported. Estimation of such errors will allow the placement of an accurate "error bar" on any simulation data reported. While it may be next to impossible to account for all of these errors this Chapter focuses on determining the error introduced by the grid (i.e. the discretization error).

In the next Chapter we also consider errors introduced by artificial dissipation.

The specific motivations of this Chapter are as follows:

1. To demonstrate typical grid resolutions needed to obtain solutions in the asymptotic range.

2. To quantify the numerical errors and uncertainties of disk drive turbulence simulations that can be computed using current computational resources.

3. To validate our computational results against experimental data sets that investigated realistic drive configurations, and finally,

4. To provide insights into certain physical aspects of the flow that may not be readily understood from experiments

In this Chapter we validate our results against the experimental data of Gross (2003), while in the next Chapter we discuss the results of Barbier (2006).

5.2   Modeling

5.2.1   Experimental Setup

The experimental setup (with which we propose to benchmark our calculations) is described in detail in Gross (2003). For clarity a schematic of the setup is shown in Figure 5.1 (reproduced directly from Gross (2003)).

The setup consists of two co-rotating glass disks of 84 mm diameter. The disk spacing is 2.0 mm and the shroud gap is 1 mm. A single e-block arm of 1.0 mm thickness was placed between them without the use of any suspension or slider assembly. The e-block arm was not actuated but it could be fixed in three positions to experiment with inner-, middle- and outer-diameter configurations. Additionally, the thickness of the e-block arm was also varied from 1.0 to 1.6 mm. A constant-temperature hot-wire anemometer was used for velocity measurements. The hot-wire probe was oriented axially at the midplane of the setup, which made it *"most sensitive to the in-plane flow speed component"*.

It is important to note that in the experimental setup the disks are shrouded for only 250 degrees of their circumferential arc-length. The remaining shroud is cut away to allow for the insertion of the e-block arm and the hot-wire probe. This region is essentially open to the atmosphere and poses some difficulty in computational modeling.

5.2.2   Computational Model Setup

Our computational model tries to closely follow the experimental setup of Gross (2003). The same geometrical dimensions are used for the disks and the e-block arm. The computational model (without the grid) is shown in Figure 5.2. An isometric wireframe view is also shown in Figure 5.3, which shows the smaller out-of-plane (z) dimension as compared to the in-plane $(r - \phi)$ dimensions.

The Reynolds number of the flow (again, tip-based) is 5,533. (Based on the outer radius it is 116,197). In any case the presence of a blunt body obstruction breaks the azimuthal symmetry and makes the flow turbulent, requiring the use of a turbulence model for simulation.

Since the computational model used here is different from those in Chapters 2 and 3, the Kolmogorov's microscale should be estimated again. This may be done using the Reynolds number associated with the largest eddies of the flow:

$$\eta = l \left(\frac{u'l}{\nu}\right)^{-3/4} \tag{5.1}$$

Here we may estimate the size of the largest eddies (l) to be equal to the disk-to-disk thickness, 2mm. And assuming that the velocity associated with the large eddies (u') is 10% of the maximum linear disk velocity, the Kolmogorov's scale (η) is approximated to be 0.0175 mm. This is a valid *apriori* assumption, based on the experimental data of Gross (2003). Also, η is in good agreement with earlier estimates of Kazemi (2004) and estimates in Chapters 2 and 3.

Given the Reynolds number of the flow, the Kolmogorov dissipation scale and the geometric volume of interest, a true direct numerical simulation would require more than 200 Million cells – which is the reason why most flows in disk drives are addressed using LES. Our simulations were performed using the CFD-ACE code utilizing the Algebraic Dynamic SGS Model (Germano et al. 1991) (i.e. ACE2 in Chapter 4).

5.2.3   Parametric Grid Generation

To study the grid dependency of our LES solutions, we conducted simulations for several different grids. To ensure a close geometric relationship between the different grids the mesh generation was parametrized. The grid was completely generated by specifying the number of nodes (and their distribution) along the edges. By changing the number of nodes uniformly (say in geometric progression) very similar (but refined) grids could be generated. The meshing strategy was to completely specify the grid parameters in the plane of the disks and then extrude the entire domain axially.

The in-plane region of the grid was divided into four distinct regions (See Figure 5.5 and 5.6):

1. Coarse structured grid: which accounts for a major part of the flow domain and does not contain any obstructions

2. Fine structured grid in the shroud gap: Here the grid is refined to resolve the streamline curvature near the shroud. However, this refinement is only sufficient to resolve the main features of the flow in the shroud, but not the boundary layer adjacent to the shroud

3. Upstream and downstream structured grid refinement: The grid is refined in the region immediately upstream and downstream of the arm. This allows the accurate

placement of the first node downstream from the solid wall of the arm. This helps
us resolve (or not resolve, depending on the grid) the separated shear layer and the
associated small turbulent structures close to the wall of a blunt body. (See Figure 5.7
for a close up view of this region)

4. Upstream and downstream unstructured grid relaxation: To interface the fine grid
   near the arm with the coarser grid in the rest of the domain an unstructured grid was
   used. The meshing tool for unstructured grids produces quadrilateral-dominant cells
   (90% quads, 10% triangular cells) which very significantly reduces the number of cells
   needed compared to a purely triangular mesh (See Figure 5.6 for close up view of this
   region)

Our initial attempts at grid dependency studies showed that the solutions changed
quite differently due to in-plane refinement as compared to out-of-plane refinement. For
this reason the grid was refined independently along the two orthogonal directions and
convergence of the solutions is reported accordingly.

5.2.4   Approximations for Boundary Conditions

The boundary conditions for the computational domain are implemented as described in
Chapter 2. However, the original experiments of Gross (2003) are "open" to the atmosphere in the region downstream of the arm. Similar boundary conditions are applied in
our computational domain by radially extending the domain 5 mm beyond the shroud.
Atmospheric pressure boundary conditions allowing the inflow and outflow of air are then
applied to the edge of this extended region. If the computational domain had not been
extended atmospheric pressure boundary conditions would have to be applied at the edge
of the rotating disk. This would, however, not be physical since we do not expect the pressure to be atmospheric immediately close to the edge of the disk. By extending the domain
outwards, a "relaxation zone" is created where the pressure in the drive may adjust to the
ambient conditions.

5.2.5   Initial Conditions and Statistical Steadiness

All of our LES calculations are initialized from steady state $k - \epsilon$ solutions to the flow
field. CFD-ACE uses the original $k - \epsilon$ implementation of Launder and Sharma (1974)
with $C_\mu = 0.09$, $C_{\epsilon_1} = 1.44$, $C_{\epsilon_2} = 1.92$, $\sigma_k = 1.0$ and $\sigma_\epsilon = 1.3$. Given the empirical
nature of the ϵ equation and the use of coefficients based on simple turbulent shear flow,
we did not expect the ϵ solution to be accurate. This is manifested in the high residuals for
the ϵ variable, which do not reduce even with a very large number of iterations (10,000).
However, velocity and pressure values at various points in the domain remained constant
(within 10% of the mean) after about 250 iterations. Using this as a guideline each $k - \epsilon$
solution was computed for 2000 iterations, and the resulting solution was used as the initial
conditions for the LES calculation.

From prior experience it is understood that instantaneous solutions of an LES are quite different (qualitatively) from the steady $k - \epsilon$ solutions. On integration in time the LES solutions change rapidly from the predicted initial conditions and gradually achieve statistical steadiness. However, since the flow is highly turbulent a local measure of steadiness (e.g. based on the convergence of the mean velocity at one point) is generally inappropriate, and a more global metric needs to be defined. For our simulations we compute the (filtered) kinetic energy and the Windage and use an energy balance argument to claim statistical steadiness.

To illustrate this technique we consider the following definitions and equations. Let u be the three-dimensional velocity vector. The kinetic energy of the flow may be defined as:

$$E(\mathbf{x}, t) = \frac{1}{2} u \cdot u \qquad (5.2)$$

while the filtered kinetic energy can be obtained by filtering the kinetic energy field,

$$\overline{E}(\mathbf{x}, t) = \frac{1}{2} \overline{u \cdot u} = E_f(\mathbf{x}, t) + k_R(\mathbf{x}, t) \qquad (5.3)$$

where the kinetic energy of the filtered velocity field is defined as,

$$E_f = \frac{1}{2} \overline{u} \cdot \overline{u} \qquad (5.4)$$

and the residual kinetic energy is defined as,

$$k_R = \frac{1}{2} \overline{u \cdot u} - \frac{1}{2} \overline{u} \cdot \overline{u} \qquad (5.5)$$

It is easy to derive the conservation equation (see Pope (2003) or Kundu (1990)) for E_f, which is (also derived as Eqn. 3.7 in Chapter 3),

$$\frac{\partial E_f}{\partial t} + \overline{u} \cdot \nabla E_f = \frac{\partial}{\partial x_i} \left\{ \overline{u_j} \left(2\nu \overline{S_{ij}} - \tau_{ij}^R - \frac{\overline{p}}{\rho} \delta_{ij} \right) \right\} - 2\nu \overline{S_{ij}} \, \overline{S_{ij}} + \tau_{ij}^R \overline{S_{ij}} \qquad (5.6)$$

where the filtered rate-of-strain tensor is given by,

$$\overline{S_{ij}} = \frac{1}{2} \left(\frac{\partial \overline{u_i}}{\partial x_j} + \frac{\partial \overline{u_j}}{\partial x_i} \right) \qquad (5.7)$$

and the sub-grid scale (residual) stress tensor τ_{ij}^R is algebraically determined from S_{ij} on applying the grid and test filters (see Sections 2.5.1- 2.5.3).

Let $\mathscr{V}(\mathbf{x})$ be the volume and $\mathscr{A}(\mathbf{x})$ be the surface area of our computational domain. $\mathscr{A}$ may be subdivided into, $\mathscr{A} = \mathscr{A}_w + \mathscr{A}_d + \mathscr{A}_o$, where the subscripts refer to stationary "walls" (both no-slip and symmetry planes), rotating "disks" and flow "outlets".

Integrating Eqn. 5.6 over $\mathscr{V}$ and converting the divergences into surface integrals over $\mathscr{A}$, we obtain the following energy balance,

$$\underbrace{\frac{\partial}{\partial t} \int_{\mathcal{V}} E_f d\mathcal{V}}_{\text{Rate of change of KE}} \quad - \underbrace{\int_{\mathcal{A}_d} \left(2\nu \overline{u_j} \; \overline{S_{ij}} - \overline{u_j} \tau_{ij}^R \right) d\mathcal{A}}_{\text{Windage}} = \qquad (5.8)$$

$$\underbrace{- \int_{\mathcal{A}_o} E_f \overline{u_j} d\mathcal{A}}_{\text{Flux of KE}} + \underbrace{\int_{\mathcal{A}_o} \left(2\nu \overline{u_j} \; \overline{S_{ij}} - \overline{u_j} \tau_{ij}^R \right) d\mathcal{A} - \int_{\mathcal{A}_o} \frac{\overline{p}}{\rho} \overline{u_j} d\mathcal{A}}_{\text{Net work by stresses at outflow}}$$

$$+ \underbrace{\int_{\mathcal{V}} \left(-2\nu \overline{S_{ij}} \; \overline{S_{ij}} + \tau_{ij}^R \overline{S_{ij}} \right) d\mathcal{V}}_{\text{Viscous and SGS dissipation}}$$

In this equation, the *Flux of kinetic energy* is the net kinetic energy produced or destroyed due to the flow of air outside our domain. The *Net work by stresses at outflow* is the work done by the surface forces (arising from the shear stress and SGS stress) on the computational volume at the boundary.

To achieve a statistical steady state it is important that an energetic balance is achieved, i.e. the energy production and dissipation balance each other, and that the net rate of change of kinetic energy be small. Since the velocity is solenoidal our calculations are mass conserving, and we do not expect very large contributions to the kinetic energy from the outflow/inflow. A dominant balance is therefore expected between the Windage and the (combined viscous and SGS) dissipation.

Based on the explanation above we computed the kinetic energy and the Windage of each simulation as the calculation progressed. Statistics of the flow (such as means, r.m.s. and higher moments) are then calculated only after the kinetic energy has "settled down", i.e. did not change by more than 5% of its mean value. This provided a systematic method for estimating statistics of the flow based on global quantities rather than on a point by point basis. The initial transients typically lasted for about 2-3 revolutions (1200-1800 time steps) of the disk. Our calculations continued until 8 revolutions – giving 6 revolutions (3600 time steps) of useful data.

5.3   Grid Dependency Studies

Grids in the 0.5 Million cell range, which may be computed on a single desktop machine, showed very poor convergence and hence the resolution was increased to approximately 2.5 Million cells. Any more refinement would have been impractical as the LES would require very long computation times. Each of the five simulations reported here was run for 2-3 weeks on a clustered Linux system using 8-32 CPUs to gather data for 8 disk revolutions. (An example of a parallel domain decomposition is shown in Figure 5.4)

The various grids used in this Chapter (labeled: Grid 1 to Grid 5) are described in Tables 5.1 and 5.2. Grids 1, 2 and 3 denote increasing z- resolution (i.e out-of plane resolution), while Grids 4, 2 and 5 represent increasing $r - \phi$ (in-plane) resolution. Since each grid was generated by completely specifying the grid in one $r - \phi$ plane, and extruding it axially, the in-plane and out-of-plane resolutions could be varied independently. The average resolution of the grid may be computed from the volume or area of the domain and the number of cells. In these tables representative grid resolutions, h, h_z and $h_{r\phi}$ are determined using the following definitions:

$$ h = \left(\frac{\text{Volume}}{N} \right)^{1/3} \tag{5.9} $$

$$ h_z = \frac{\text{Axial dimension}}{N_z} \tag{5.10} $$

$$ h_{r\phi} = \left(\frac{\text{In-plane area}}{N_{r\phi}} \right)^{1/2} \tag{5.11} $$

In presenting the convergence results for two orthogonal directions we often notice that the two sets of grids (1-2-3 and 4-2-5) are converging to different results when extrapolated to $h = 0$. Nevertheless, the actual value of the result at $h = 0$ is not of much consequence, since it is significantly affected by several factors other than the grid (as discussed briefly earlier). However, extrapolated error and the grid convergence index (GCI) are very useful in quantifying the uncertainty of the results.

5.3.1   Kinetic Energy and Windage

We start by discussing the convergence of global quantities such as the kinetic energy, windage and drag on the arm. The windage and the drag on the arm are especially important to the disk drive community, because they refer to the power required by the motor to run the disks and the force on the actuator, respectively. These quantities are referred to as "global" because they are obtained by integration in space, and the integrand is dependant on the properties of the flow at several locations. Global quantities are expected to show better behavior in convergence than local estimates, since the integration should smooth out local errors and present an average estimate of the rates of convergence.

The evolution of the kinetic energy and windage are shown in Figure 5.8. In this Figure results are presented for the finest grid (Grid 5) and error bars are included based on the Grid Convergence Index (GCI) of the mean kinetic energy and windage. In obtaining the GCI we have followed the guidelines of the 'ASME Journal of Fluids Engineering policy statement on the control of numerical accuracy'. In this figure and all subsequent figures of this Chapter error bars are applied to data from the finest grid itself, instead of the more customary practice of using the extrapolated data. Usually the extrapolated solutions are close enough to the finest grid calculations to be included in the uncertainty error bars. Nonetheless, if these computations were to serve as a benchmark for future validation efforts the fine grid data would be more useful than the extrapolated solutions.

In Figure 5.8 the quantities are non-dimensionalized using the following definitions: Let, $U_o = \Omega R_o$ be the disk edge velocity, where Ω is the rotation speed and R_o is the disk outer radius. Let $\mathbb{V}$ be the volume of the domain and the $\mathbb{A}_d$ be the area of the disks. Then the non-dimensional kinetic energy and windage may be defined as,

$$k^* = \frac{\frac{1}{2}\int_{\mathscr{V}} \mathbf{u} \cdot \mathbf{u} d\mathscr{V}}{\frac{1}{2}U_o^2 \mathbb{V}} \tag{5.12}$$

$$W^* = \frac{\int_{\mathscr{A}_d} \left(2\nu\overline{u_j S_{ij}} - \overline{u_j}\tau_{ij}^r\right) d\mathscr{A}}{\left[\frac{1}{2}U_o^2\right][U_o] \mathbb{A}_d} \tag{5.13}$$

In the same Figure 5.8 the third sub-figure shows the rate of change of k^*. Finally, in the fourth sub-figure the difference between the rate of change of kinetic energy and windage (i.e. right-hand-side of Eqn 5.8) is plotted, which is mainly the combined viscous, SGS and numerical dissipation.

From the figure we notice that the rate of change of kinetic energy decreases almost to zero after about 2 revolutions of the disk. The kinetic energy *decays* from its steady value, indicating the $k-\epsilon$ solutions tend to overpredict the kinetic energy of the flow. Interestingly, the decay in kinetic energy is very close to an exponential function, and a direct comparison of an exponential curve with the kinetic energy is plotted in Figure 5.9. The rate of decay was found to have a time constant of 0.737 revolutions, suggesting that the kinetic energy will achieve 5% of it's mean value in 2.209 revolutions. In reporting the rest of our results our statistical averaging is started after the kinetic energy is within 5% of its converged mean value. In Figure 5.8 this is a little after 2 revolutions. At about 3 revolutions the change in kinetic energy is less than 1% of the mean.

The error bars in Figure 5.8 are based on the data from Table 5.3 by using the higher value of GCI. The table also reports the absolute error in the solutions (e_a) and the error in the extrapolated solution (e_{exp}). The GCI is computed separately in the in-plane and the axial directions and is reported in Table 5.3. Since we are dealing with global quantities the GCI calculations are based on the global grid size h and not on directional resolutions such as h_z and $h_{r\phi}$. The kinetic energy and windage both show monotonic convergence in both the z- and $r - \phi$ directions. In Table 5.3 the calculated order of convergence ranges from 1.02 to 2.71, which is in good agreement with the formal order of accuracy, 2. This is also an indication that the chosen grids are in the asymptotic range. In general, increasing the resolution causes both the mean kinetic energy and windage to decrease. From this one may infer that increasing the number of cells allows the resolution of smaller flow structures associated with smaller kinetic energies. The energy cascade from the larger to the smaller eddies is thus responsible for lowering the total kinetic energy of the domain. Interestingly, in case of both the kinetic energy and windage we observe a higher sensitivity to the z-resolution than the $r - \phi$ resolution. This also leads to the result that $GCI_z^{23} > GCI_{r\phi}^{25}$, which implies a higher uncertainty due to the resolution in the z direction. From Chapter 2 it it known that the velocity profiles in a disk drive are similar to turbulent Couette flow. Since the principal mechanism for generating kinetic energy is from the rotating disks the interdisk resolution plays a vital role in the kinetic energy of the flow. The momentum being "pumped" into the domain is highly dependant on the resolution in the boundary layer. On the other hand, the in-plane resolution (especially in the wake of the arm) determines the rate of loss of kinetic energy to the viscous and SGS sinks. The overall result is that the energetics of the flow domain are more sensitive to the z-resolution than the $r - \phi$ resolution in the range considered.

5.3.2 Off-Track and On-Track Drag

A similar time history of the coefficient of drag, C_D, on the actuator is plotted in Figures 5.10 and 5.11. The time history is shown for the final 6 revolutions of the computation. The coefficients are further decomposed into Off-Track and On-track directions, where again, off-Track is the direction perpendicular to the axis of the e-block arm, and On-track is the direction parallel to the axis of the e-block arm. In computing these coefficients the projected areas of the arm and the disk edge speed U_o are used.

From Figure 5.10 we observe that C_D Off-track is almost twice as large as C_D On-track, which is due to the orientation of the arm in the rotating flow. Figure 5.10 also shows that increasing the z-resolution increases the mean Off-Track drag but decreases the mean On-track drag. Interestingly, the RMS values of both the Off-Track and On-track drag reduce. This suggests that under-resolved simulations, which are dominated by the large scale motions, tend to over predict the fluctuations of pressure acting on the arm. Increasing the resolution allows the cascade to (slightly) smaller scales than before, resulting in smaller fluctuations at the large eddy level.

There is little difference in the convergence results for the z and $r - \phi$ directions for the drag, given in Table 5.3. Our results indicate that the GCI is high (20-30%) for the mean and RMS values of drag coefficients. The RMS values of the drag coefficients show oscillatory convergence in the $r - \phi$ direction and hence the GCI is not reported.

The RMS values of the drag coefficient on the arm may be broken down into frequency components using Parsevals theorem. Information regarding the amount of energy associated with different frequency bands is important to the disk drive component designers, who may then design structures with natural frequencies that do not fall in the heavily excited bands.

Figures 5.12 and 5.13 show the RMS contribution from different frequency bands to the Off-Track drag coefficient. Similarly, Figures 5.14 and 5.15 show the RMS contribution from different frequency bands to the On-Track drag coefficient. Interestingly, some clear trends are demonstrated: By increasing the resolution, the low frequency contribution (0-1 kHz) decreases, while the higher frequency contribution, especially 1-6 kHz, increases. This trend is consistently demonstrated in both the z- and $r - \phi$ directions; however, as seen in Figures 5.12 and 5.13, convergence is monotonic in z- but oscillatory in the $r - \phi$ direction. We note that to obtain the resultant RMS due to all frequency bands algebraic addition is not permitted, but the RMS values should be added geometrically: by summing their squares and taking the square root.

The Figures 5.12 to 5.15 display an important trend in the frequency components of the excitation force on the actuator. This data (in the monotonically convergent cases) may be used to obtain the extrapolated solution and the GCI. It is most useful to directly compare the extrapolated values with the values from the finest grid (Grid 3), along with the GCI. This is done in Figure 5.16 for the Off-Track component and Figure 5.17 for the On-Track component. Again, the figures show very interesting results. Firstly, the difference between the extrapolated solution and the solution from the finest grid decreases with increasing frequency. Generally, there is excess energy in the lower frequencies, but less energies in the higher frequencies. Secondly, the GCI decreases with increasing frequency, indicating that the LES solutions converge much faster in the higher frequency components. In Figure 5.16 the very high GCI value in the 6-10 kHz range is hard to explain and may be considered spurious. The contributions in the 10-50 kHz range are not analyzed for convergence since the values are very close to each other.

5.4   Experimental Validation

In this section our numerical results are directly compared with the experimental data of Gross (2003). Two experimental data sets are available: Shown in Figure 5.18 are measurement locations along a single line in the wake of the arm, and shown in Figure 5.19 is the measurement grid in a broader rectangular area, again downstream of the e-block arm. The measurement area in Figure 5.19 is referenced by an x-y coordinate system

5.4.1   Measurements along a line

Figure 5.20 shows the mean velocity along the measurement line, Figure 5.21 shows the RMS velocity and Figure 5.22 shows the turbulence intensity (i.e. the ratio of the RMS to the mean velocity). In these figures the distance along the measurement line is non-dimensionalized by the length of the line, so all of the plots range from to 0 to 1. The 0 end of the plot corresponds to outside edge of the disk, while the 1 end of the plot corresponds to the inner location (see Figure 5.18) In all of the figures the percentage occurrence of oscillatory convergence is displayed at the top along with the average order of convergence. In plotting the error bars on the figures the GCI was determined using the usual formula, but with the average order p_{avg}, of the method. The error bars were then included in the figures at ten equispaced locations. In all three figures a higher number of points showed oscillatory convergence in the $r-\phi$ direction than in the z-direction, hence the GCI estimates are from convergence in the z-direction.

In general there is higher agreement in the mean quantities than in the RMS quantities. The spatial variation of the mean velocity along the measurement line is in fairly good agreement with the experimental data. Remarkably, the agreement is very good close to the outer edge of the disk, where we expect the influence of the outflow boundary condition. This indicates that the relaxation region included in our simulations provides a good estimation to the physical outflow boundary. The velocity profiles show a higher local variation compared to the experimental data – which is smoother. The reason for this may be that the experimental data is based on readings taken over several minutes (i.e. several thousand revolutions) while the computational data is averaged for 6 revolutions only. In general, the percentage of points showing oscillatory convergence is higher for the RMS velocity than the mean velocity.

It is a general observation that the LES results tend to over predict the RMS fluctuations of velocity. This is consistent with the drag results outlined previously, and it tends to corroborate the notion that LES simulations on the current grids tend to under resolve the smaller scales of motion, leading to higher fluctuations in the large scales. Figure 5.22 shows the turbulence intensity along the measurement line, which is the ratio of the RMS to the mean velocities. Again, the turbulence intensity is higher in the simulation as compared to the experiment, but there is good agreement close to the outflow boundary condition. In Gross (2003), in addition to the turbulence intensity, the mean and RMS dynamic pressure head is also reported. These quantities can be easily deduced from the mean and RMS velocities, hence we do not report them here.

In Figure 5.23 the frequency spectrum of the velocity fluctuations is plotted, which may be compared with the experimental results in Figure 5.24. Several observations can be made with regard to Figure 5.23. Firstly, the data is more noisy than the experimental results because of the limited data set available. Secondly, there are no clear peaks corresponding to frequency locking. The orientation of the arm in the rotating flow and the complex geometry of the arm itself, generated a complex wake. The vorticity shed from the arm organizes itself into eddies behind the arm but this phenomenon is not self-selective of any frequency. The fluctuations are contained in the low frequencies (0-6 kHz) and are much smaller at frequencies beyond that. This unsteadiness appears to be mostly random, but the flow structures that are shed are long lived and coherent. These flow structures are carried around by the rotating disk and are dissipated in time.

5.4.2 Measurements on the area

Figures 5.25 and 5.26 show a direct comparison between the LES and experimental data. Again, the LES results are from Grid 3, and the percentage oscillatory convergence and average order of accuracy are included on the top of each sub-figure. The error bars on the LES data are based on the GCI from the average order of accuracy.

In the Figures 5.25 and 5.26 the mean and RMS velocities are plotted as functions of the y-coordinate (ranging from -6 to 6). Different x-locations (ranging from 2 to 6) are plotted in different sub-figures. See Figure 5.19 for the location of the x-y coordinate system.

As shown in these figures, there is better agreement in the mean velocities than in the RMS velocities. The experimental data shows that the mean velocity has a radial gradient and there is a well defined transition from a smaller velocity to a larger velocity when going from $y = -6$ to 6. This is because the flow is blocked immediately downstream of the arm, and is accelerated in the space between the arm and the hub. The LES data also shows a similar trend, but the transition is a little further away from the hub. For $x = 12$ and $x = 10$ the magnitudes are in remarkable agreement.

In terms of RMS the experiments show a moderate level of fluctuations in the wake, and a slight increase in the fluctuations in the region where the mean velocity transitions, followed by much smaller fluctuations approaching the hub. The LES results, however, show different qualitative features. They exhibit a higher level of RMS fluctuations and a significantly higher peak in the flow transition region. Additionally, $x = 12$, $x = 10$ and $x = 8$ also show peaks in RMS near the hub.

Finally, Figures 5.27 to 5.30 graphically summarize the results in the rectangular measurement area. While Figures 5.27 and 5.28 share the same color scale for the mean velocity, Figures 5.29 and 5.30 have different color scales for the RMS as denoted. The figures for mean velocity show the transition of the velocity from the blocked region to the accelerated region. The experimental figure also shows, by a dotted line, the location of the suspension slider assembly if it were to be included in the setup. Figures 5.29 and 5.30 show the larger differences between the RMS fluctuations, as discussed earlier.

5.4.3 Frequency contribution to RMS

The frequency contribution to the RMS from different frequency bands is now discussed in detail. Figures 5.31 and 5.32 compare the 0-2 kHz frequency contributions to the RMS, while Figures 5.33 and 5.34 compare the contributions from 2 to 20 kHz.

In both cases the RMS from the LES is approximately two to three times larger than that measured in the experiments. While it is unclear what the exact source of the discrepancy is, it is well known that simulations tend to over predict some components of the RMS fluctuations. As seen in Chapter 4, the flow across a square cylinder was computed and streamwise Reynolds stresses (i.e. the streamwise velocity RMS fluctuations) were over predicted. This over prediction was due to the nature of the SGS model itself and we may conclude that modeling error contributes significantly to the prediction of the velocity fluctuations.

Both Figures 5.32 and 5.34 show the clear stream of shed eddies that contribute to higher fluctuations. The region blocked by the arm has higher fluctuations than the accelerated flow region. The thesis of Gross (2003) also breaks down the 2-20 kHz contribution to the RMS in to 2-6, 6-10 and 10-20 kHz bands. Figures 5.35 through 5.40 provide a direct comparison between the LES and the experiments for these frequency bands. The general trend is that the LES consistently predicts higher fluctuations compared to the experimental data in all frequency bands. With increasing grid resolution, both in the $r - \phi$ and z-directions, the 2-6 and 6-10 kHz contribution to the RMS increases, while the 0-2 kHz contribution decreases. This trend is exactly similar to the trend in the drag coefficients shown in Figures 5.12-5.15 and is hence not repeated. This leads to the conclusion that when performing calculations on successively refined grids LES solutions converge to spectral contents that do not quantitatively agree with the experimental spectra. Thus grid-free LES solutions can never agree perfectly with experiments, which is most likely due to the deficiency in the SGS model, as suggested earlier. Other factors such as limited data for averaging LES solutions, influence of boundary conditions and the uncertainty in the hot-wire measurement process may also contribute to the discrepancy between the results.

5.4.4 Length and time scales

The characterization of the turbulent flow is not complete without the specification of a time scale and a length scale. The integral time scale of the flow may be computed using the normalized auto-correlation function,

$$\rho(s) = \frac{\frac{1}{T} \int_0^T u'_\phi(t) u'_\phi(t+s) dt}{\frac{1}{T} \int_0^T u'^2_\phi(t) dt} \tag{5.14}$$

where $u'_\phi = u_\phi - \langle \overline{u_\phi} \rangle$, according to the Reynolds decomposition.

The integral time scale may be then computed as,

$$\tau = \int_0^\infty \rho(s) ds \tag{5.15}$$

and invoking Taylor's frozen field hypothesis (see Pope 2003) the integral length scale may be computed as:

$$\lambda = \tau \langle \overline{u_\phi} \rangle \tag{5.16}$$

The integral time scale and length scale are shown in Figures 5.41 and 5.42. The time scale is non-dimensionalized to represent the number of disk rotations. The time scale is the largest in the accelerated part of the area and is relatively small in the region of the wake. This indicates that although the flow is being accelerated in this region, the flow remains largely laminar and fluctuations are well correlated for almost a whole revolution of the disk. In the more turbulent wake, the fluctuations flow remain uncorrelated, and the integral time scales are small. In the laminar region the combined effect of flow acceleration and larger time scales leads to much larger length scales. The length scales are much smaller in the wake. This indicates that the largest flow structures in the domain are contained in the laminar flow region, and the wake is characterized by much smaller flow structures with shorter life spans.

We also note that most of the eddies in the turbulent wake have a length scale of about 2 mm or less, which is also the disk-to-disk spacing in the model. Hence, estimations of the Kolmogorov's microscale based on this estimate is valid, as done previously.

Finally, the cross term of the (time averaged) Reynolds stress tensor representing $\langle \overline{u_r u_\phi} \rangle$ is plotted in Figure 5.43. It clearly shows a distinct ridge in the stress component in the region where the flow transitions from the accelerated region to the wake region. This indicates the region of strong production of turbulence and the region where the turbulent field is anisotropic. The negative sign of the stress component is typical of a turbulent shear flow and indicates the production of turbulence from the interaction between the fluctuating field and the mean field.

5.5 Conclusions

This Chapter has dealt with two crucial aspects of any simulation activity: grid dependency studies and experimental validation. Further comparisons with experiments (that use a different experimental technique) are done in the next Chapter. To summarize the main conclusions:

1. Comprehensive grid convergence results have been carried out for flows in hard disk drives. We found that grids in the 2-2.5 Million cells range (for a 3 inch drive) are in the asymptotic range. While it is customary to vary the grid uniformly in all three dimensions and report the convergence, such an effort would have missed the independent sensitivity and convergence characteristics of the in-plane and out-of-plane resolutions.

2. In the face of limited computational resources and very long simulation time we have also outlined a rigorous and novel technique to define the (statistical) steadiness of the flow. This is based on monitoring the kinetic energy and windage of the flow. We found that simulations initiated from steady $k - \epsilon$ solutions decay exponentially to their steady values, which is helpful in deciding an averaging interval for reporting the statistics of the flow.

3. Our grid convergence results mainly show monotonic convergence in z and oscillatory convergence in $r - \phi$, with GCI values approximately 20-30 % for most quantities. More importantly, we noticed higher sensitivity of the quantities to the z-resolution, which indicates the importance of resolving the axial dimension adequately for accurate simulation. The simulation results also show that increasing the grid resolution changes the spectral content of the drag on the arm. Increasing grid resolution decreases the 0-1 kHz content while increasing the higher 2-6 kHz spectral content. Finally, the results of this Chapter can assist disk drive CFD practitioners to estimate the grid based uncertainty of their simulations and compensate (correct) their results based on the data presented here.

4. In validating our LES results with the hot wire experiments we found good agreement in the mean quantities but larger discrepancies in the RMS quantities. Generally, statistical quantities reported in an LES do not account for the direct influence of the unresolved scales and hence such comparisons should be made with caution. Our findings show that LES results tend to over predict the fluctuations in almost all frequency bands, and the spectra converge to solutions that do not quantitatively agree with the experiments. While the contribution from the highest frequencies is very small, (e.g. the contribution of 10-20 kHz range to the RMS is only 0.6%) LES results still overpredict the amount of fluctuations arising from this frequency band. We may attribute this discrepancy to both the simulations and experiments – LES has a tendency to overpredict streamwise fluctuations and underpredict cross-stream fluctuations (directly from Chapter 4), while hot-wire has a tendency to underpredict streamwise fluctuations because the hot wire probe has a finite length (0.2 mm, 10% of the disk-to-disk thickness) and averages the flow velocity over that distance.

The next Chapter is roughly a continuation of this experimental validation effort. However, instead of performing grid convergence studies again, we test other matters such as artificial dissipation.

5.6 Tables

Table 5.1: Grids with variable out-of-plane (z) resolution

Grid Name	Volume mm^3	Number of cells, N	Number of out-of-plane cells, N_z	Number of in-plane cells, $N_{r\phi}$	h mm	h_z mm	$h_{r\phi}$ mm
Grid 1	1.043×10^4	1,101,264	16	68,829	0.2116	0.1250	0.2735
Grid 2	1.043×10^4	1,651,896	24	68,829	0.1848	0.0833	0.2735
Grid 3	1.043×10^4	2,202,528	32	68,829	0.1679	0.0625	0.2735

Table 5.2: Grids with variable in-plane $(r\phi)$ resolution

Grid Name	Volume mm^3	Number of cells, N	Number of out-of-plane cells, N_z	Number of in-plane cells, $N_{r\phi}$	h mm	h_z mm	$h_{r\phi}$ mm
Grid 4	1.043×10^4	1,171,632	24	48,818	0.2072	0.0833	0.3248
Grid 2	1.043×10^4	1,651,896	24	68,829	0.1848	0.0833	0.2735
Grid 5	1.043×10^4	2,361,324	24	98,389	0.1640	0.0833	0.2288

Table 5.3: Grid Convergence results for global quantities

Global Quantity		p_z	$e_{a,z}^{23}$	$e_{ext,z}^{23}$	GCI_z^{23}	$p_{r\phi}$	$e_{a,r\phi}^{25}$	$e_{ext,r\phi}^{25}$	$GCI_{r\phi}^{25}$
			(%)	(%)	(%)		(%)	(%)	(%)
Kinetic Energy	Mean	1.85	2.66	15.87	17.12	2.71	2.19	6.09	7.17
Windage	Mean	1.45	1.19	8.71	10.01	1.19	1.19	8.54	9.83
Off-track drag	Mean	1.56	2.48	13.28	19.15	1.02	2.39	15.63	23.16
	RMS	1.09	1.26	12.85	14.23	–			
On-track drag	Mean	1.74	1.90	11.69	13.09	1.48	5.02	34.93	32.36
	RMS	1.74	2.38	15.05	16.35	–			

5.7   Figures

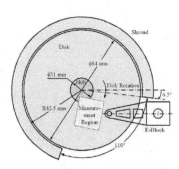

Figure 5.1: Experimental Setup (from Gross 2003) (diagram is not to scale)

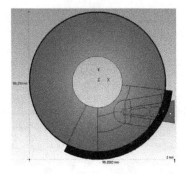

Figure 5.2: Top View of Computational Model (diagram is to scale)

Figure 5.3: Wireframe Isometric View of Computational Model

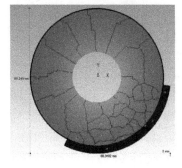

Figure 5.4: Typical domain decomposition for parallel computation

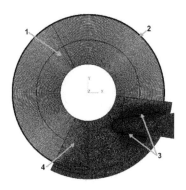

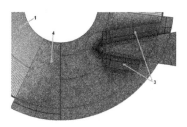

Figure 5.5: Top view of the computational grid (Grid 1 in Table 5.1)

Figure 5.6: Closeup showing regions 1, 3 and 4 (Grid 1 in Table 5.1)

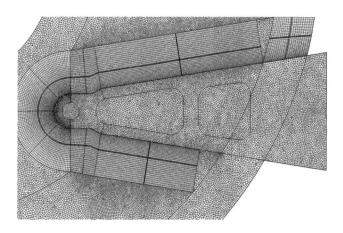

Figure 5.7: Closeup showing region 3 (Grid 1 in Table 5.1)

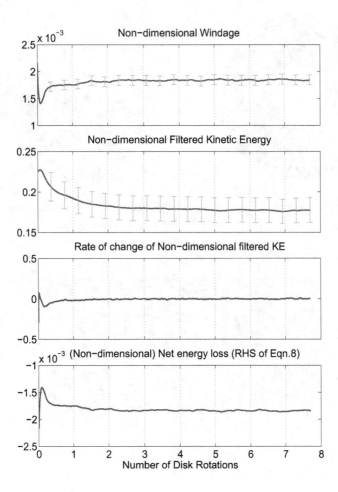

Figure 5.8: Global energy-related quantities for Grid 5

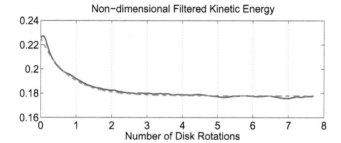

Figure 5.9: Global energy-related quantities for Grid 5

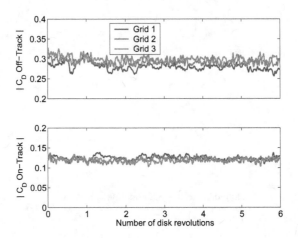

Figure 5.10: Time history of the drag coefficients for Grids 1, 2 and 3.

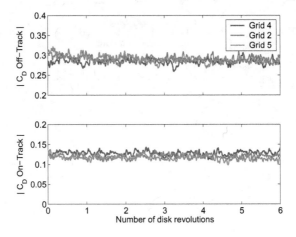

Figure 5.11: Time history of the drag coefficients for Grids 4, 2 and 5.

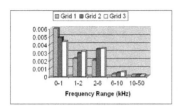

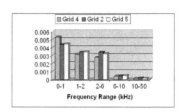

Figure 5.12: Contribution to C_D Off Track RMS from different frequency bands, shown for Grids 1, 2 and 3

Figure 5.13: Contribution to C_D Off Track RMS from different frequency bands, shown for Grids 4, 2 and 5

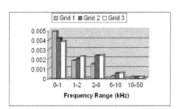

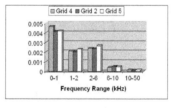

Figure 5.14: Contribution to C_D On Track RMS from different frequency bands, shown for Grids 1, 2 and 3

Figure 5.15: Contribution to C_D On Track RMS from different frequency bands, shown for Grids 4, 2 and 5

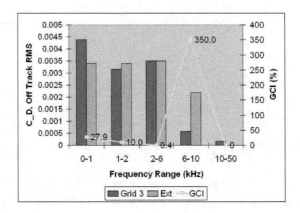

Figure 5.16: Comparison of the RMS contributions to C_D Off Track from the finest grid and the extrapolated contributions. Also shown is the GCI_z^{23} across different frequency bands

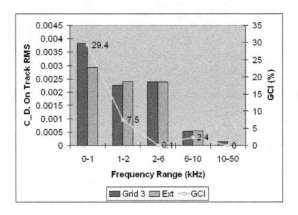

Figure 5.17: Comparison of the RMS contributions to C_D On Track from the finest grid and the extrapolated contributions. Also shown is the GCI_z^{23} across different frequency bands

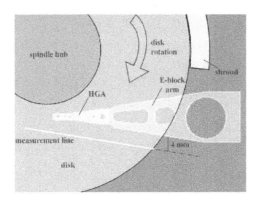

Figure 5.18: Location of measurement **line** for hot-wire experimental data, from Gross (2003)

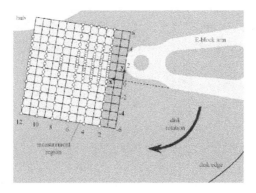

Figure 5.19: Location of measurement **area** for hot-wire experimental data, from Gross (2003)

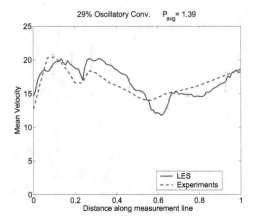

Figure 5.20: Mean flow speed along measurement line. LES data is plotted along with error bars representative of GCI

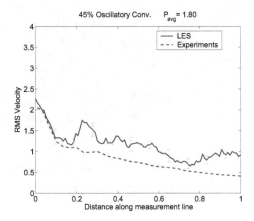

Figure 5.21: RMS of flow speed fluctuation along measurement line. LES data is plotted along with error bars representative of GCI

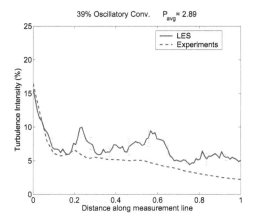

Figure 5.22: Turbulence intensity along measurement line. LES data is plotted along with error bars representative of GCI

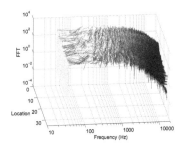

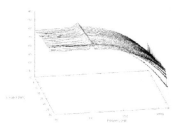

Figure 5.23: Frequency spectrum of velocity fluctuations at different locations, from Grid 5

Figure 5.24: Frequency spectrum of velocity fluctuations at different measurement locations, reproduced from Gross (2003)

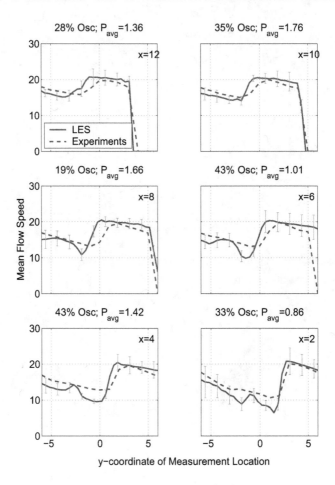

Figure 5.25: Mean flow speed in m/s, plotted for various x-positions. Both Experimental and LES data is shown, the latter with error-bars based on the GCI. At the top of each sub-figure, the percentage occurrence of oscillatory convergence and the average order of convergence is printed.

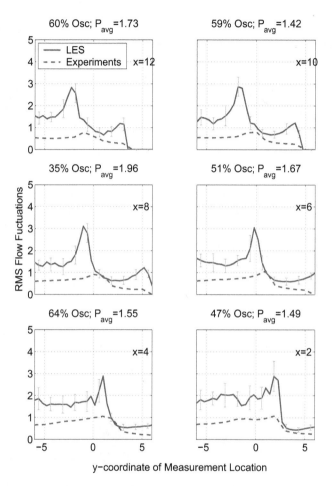

Figure 5.26: RMS flow fluctuations in m/s, plotted for various x-positions. Both Experimental and LES data is shown, the latter with error-bars based on the GCI. At the top of each sub-figure, the percentage occurrence of oscillatory convergence and the average order of convergence is printed.

Figure 5.27: Mean flow velocity over entire measurement area, from LES

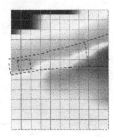

Figure 5.28: Mean flow velocity over entire measurement area, from experiments of Gross (2003)

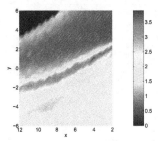

Figure 5.29: RMS flow fluctuations over entire measurement area, from LES

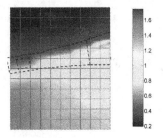

Figure 5.30: RMS flow fluctuations over entire measurement area, from experiments of Gross (2003)

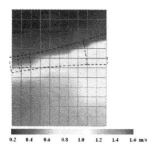

Figure 5.31: 0-2 kHz contribution to RMS flow fluctuations, from Gross (2003)

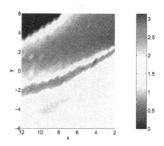

Figure 5.32: 0-2 kHz contribution to RMS flow fluctuations, from LES

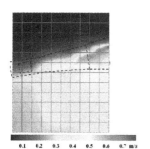

Figure 5.33: 2-20 kHz contribution to RMS flow fluctuations, from Gross (2003)

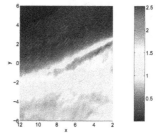

Figure 5.34: 2-20 kHz contribution to RMS flow fluctuations, from LES

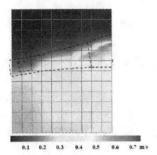

Figure 5.35: 2-6 kHz contribution to RMS flow fluctuations, from Gross (2003)

Figure 5.36: 2-6 kHz contribution to RMS flow fluctuations, from LES

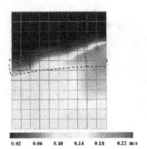

Figure 5.37: 6-10 kHz contribution to RMS flow fluctuations, from Gross (2003)

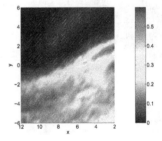

Figure 5.38: 6-10 kHz contribution to RMS flow fluctuations, from LES

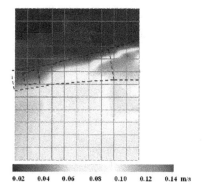

Figure 5.39: 10-20 kHz contribution to RMS
flow fluctuations, from Gross (2003)

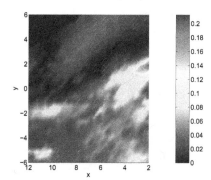

Figure 5.40: 10-20 kHz contribution to RMS
flow fluctuations, from LES

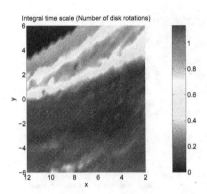

Figure 5.41: Integral time scale of the flow (in number of disk rotations)

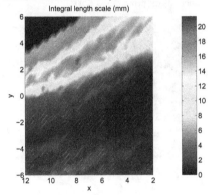

Figure 5.42: Integral length scale of the flow (mm)

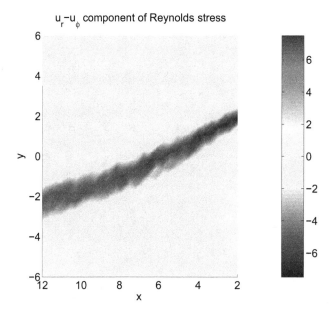

Figure 5.43: Cross term of the time-averaged Reynolds stress tensor

Chapter 6

A Comparison with Experiments

of Barbier (2006)

6.1 Introduction

In the previous Chapter we performed extensive grid convergence studies and validated our simulation data with the hot-wire experiments of Gross (2003). The general conclusions were that the uncertainty of the experimental and computational mean flow velocities lie well within each others range leading to (loosely defined) validation. On the other hand, the RMS fluctuations of the simulation were larger than the experiments, sometimes by 2-3 times. We attributed this discrepancy to both the simulations and experiments – LES has a tendency to overpredict streamwise fluctuations and underpredict cross-stream fluctuations, while hot-wire measurements have a tendency to underpredict streamwise fluctuations due to the finite length of the probe. Finally, a major contribution of the Chapter is in understanding the uncertainty of our simulation work due to the discretization grid.

In this Chapter we compare our simulation results with those from the thesis of Barbier (2006). These measurements were made primarily using particle-image-velocimetry (PIV), which is fundamentally different from the hot-wire based experiments of the previous Chapter. We would like to investigate whether better agreement can be found between CFD and PIV as compared to CFD and hot-wire anemometry. Moreover, the experiments of Barbier (2006) are the only other public-domain datasets available for validation – hence to seek completeness in this dissertation it is a useful exercise to compare our results with theirs.

Ideally such validation efforts should be done in the context of grid convergence studies, as was done in the last Chapter. However, grid convergence studies for the current case will not lead to any significantly new results. Instead, the closed computational domain allows the (approximate) quantification of numerical dissipation, by a novel technique. This is explained in Sec 6.4.

6.2   Problem Setup

6.2.1   Experimental Setup

A photograph showing the experimental setup of Barbier (2006) is shown in Figure 6.1. It consists of a twice scaled up (2×) model of a realistic disk drive enclosure. Scaling up the model allows for smaller rotational speeds in the drive at the same Reynolds number. The setup consists of two co-rotating disks of 200 mm diameter rotating at speeds between 250 to 3000 RPM. The disk to disk spacing was 4.8 mm. The disks were fully shrouded, except at the location where the arm is inserted. The disk to shroud gap was 2mm. A simplified e-block arm and a pair of suspensions form the actuator. The suspensions have a pair of non-flying sliders attached to their ends. The gap between the slider and the disk was 0.075 mm – however, this gap was not modeled in the simulations. Instead, the disk to disk spacing was reduced from 4.8 mm to 4.65 mm. The e-block arm was 2.95 mm thick, while the suspension was 0.75 mm thick. The overall length of the actuator was 123.8 mm.

For the PIV experiments particles were seeded from a hole near the base of the actuator (shown in Figure 6.1). Measurements based on the laser sheet were made at several locations upstream and downstream of the actuator arm.

6.2.2   Computational Model

Figure 6.2 depicts a schematic of the experimental setup (to the left) along side the computational model (center). To the right is the domain decomposition used for parallel computation (using 32 CPUs). The computational model was an exact geometric replica of the experimental setup. However, the usual simplifying assumptions still apply:

- The shroud gap was modeled as a symmetry plane

- The disks were modeled as rigid rotating walls, with no vibrations or run-outs

- The actuator did not feed back vibrations to the flow

- The sliders were simply supported on the disks, and the fly height gap was not modeled

Two LES calculations were performed. Relevant details of these simulations are given in Table 6.1. To make a comparison between the experiments and the simulations we chose two rotational speeds: 3000 RPM, which is the extreme case (i.e. the largest Reynolds number investigated), and 1500 RPM. Given the large size of the computational domain (for the 2× model) the average resolution of these simulations are less than those from the previous Chapter. The finest (average) resolution in Chapter 5 was about 0.165 mm which is higher than the current resolution by 2-3 times. For this reason, we expect larger grid-based uncertainty in our results. From Table 6.1 one can readily observe that such computations take several weeks to run. Increasing the grid size to be able to resolve finer details would increase the computational cost exponentially.

6.3 Results

In this section two types of results are discussed: velocity profiles at the midplane between the disks, both upstream and downstream of the arm (for 3000 RPM and 1500 RPM), and frequency spectra at a few locations in the domain (only for the 1500 RPM case). Experimental data is available for comparison in the former case, while in the latter, spectra of the turbulent kinetic energy are used to characterize the turbulence in the flow.

6.3.1 Velocity profiles

The thesis of Barbier (2006) presents PIV data along various chords in the domain. Generally these chords are referenced with respect to an origin (either O_1 or O_3). We begin the discussion by first considering the upstream locations, and then we consider the downstream ones.

Figure 6.7 shows several quantities plotted along chord O_1-Up which is the most **Up**stream measurement available. The location of this chord is shown in the bottom left corner of the figure. In this same figure, five quantities are plotted (and labeled accordingly): mean and RMS of the radial and azimuthal velocities and the time averaged cross term of the Reynolds stress tensor ($\langle \overline{u_r}\,\overline{u_\phi} \rangle$). This term is also the covariance between $\overline{u_r}(t)$ and $\overline{u_\phi}(t)$. The mean and RMS velocities are non-dimensionalized by the disk edge velocity, while the covariance term is non-dimensionalized by the square of the disk edge velocity.

Figure 6.7 (chord O_1-Up), Figure 6.8 (chord $O_1 - 1$) and Figure 6.9 (chord $O_1 - 2$) depict the above quantities upstream of the arm for the 3000 RPM case. Several general conclusions can be made: We find that there is relatively good agreement in the mean azimuthal and radial velocities. The agreement is higher farther upstream from the arm, and decreases as the location comes closer to the arm. Also the agreement of the azimuthal velocity is better than that of the radial velocity – and this is expected since the rotating azimuthal flow is the primary streamwise flow, while the radial flow is a secondary cross stream flow.

In Figure 6.7 (chord O_1-Up) we see that the simulation azimuthal velocity goes negative and asymptotes to zero near the hub. This is an indication of a recirculation zone (also called circulation bubble), formed due to the adverse pressure gradient due to the arm. However the experiments do not show this feature, and the velocity smoothly asymptotes to the velocity of the inner hub. When going from chord O_1-Up to chord $O_1 - 1$ this recirculation zone disappears in the simulations, and the azimuthal velocities agree better along chord $O_1 - 1$.

Generally our simulations tend to predict higher RMS fluctuations when compared to the PIV experiments, as they did in the comparison with the hot wire experiments of the previous Chapter. In any case there is good agreement of the spatial profiles along all three of the chords considered. The cross term of the Reynolds stress (as seen in Chapter 4) is a measure of the anisotropy of the flow. Again, we find reasonable agreement in the profiles of the experiments and the simulations. At any given location the magnitudes may relatively differ by large amounts between the simulations and the experiments, but the general locations of the peaks are well reproduced.

Figures 6.10– 6.12 depict velocity data along three chords with O_3 as the origin. Even though O_3 is the origin of the three chords the azimuthal and radial velocities are still in the global coordinates of the disks (i.e. with O_1 as the origin). Figure 6.10 shows data along chord $O_3 - 1$, which is upstream of the arm. This figure is consistent with the previous figures in that it shows reasonable agreement upstream of the arm. However, close to O_3 there is a larger discrepancy in the mean azimuthal velocity between the experiments and simulation. Chords $O_3 - 2$ and $O_3 - 3$ (Figures 6.11- 6.12) lie downstream of the arm, and in both these cases we observe a large discrepancy between the mean azimuthal velocities. Along chord $O_3 - 2$ there is reasonable agreement between the RMS profiles, but this is not the case at Chord $O_3 - 3$.

This implies that considerable differences in the velocities are predicted by the simulation versus those predicted by the experiments, in the wake of the arm. Several reasons could explain these discrepancies. On the computational side insufficient grid resolution in the wake could be a potential reason. Since velocity profiles match well upstream, but do not match well downstream, the grid resolution in the wake may not be fine enough to resolve the smaller features of the flow. On the experimental side it is well known (and acknowledged in Barbier (2006)) that due to refraction one cannot maintain a very thin sheet of laser for PIV measurement. This suggests that the PIV technique may actually integrate the velocity over a thickness range covering the midplane. And since velocities are higher closer to the disks this integrated mean velocity is likely to larger than the mean velocity at the exact mid plane.

Figures 6.13, 6.14 and 6.15 provide data along chords $O_3 - A$, $O_3 - B$ and $O_3 - C$, respectively, for the 1500 RPM simulation. In this simulation the Reynolds number is halved as compared to the previous simulation, while the grid resolution is increased by approximately 20%. For this reason we find very good agreement between the simulations and the experiments at the upstream chords $O_3 - A$ and $O_3 - B$ in all the quantities plotted. There is good agreement in the magnitudes and signs and also the locations of peaks in the RMS velocities. However, chord $O_3 - C$ still shows a mismatch between the numerical and experimental values. The agreement in the Reynolds stresses (and the RMS velocities) is improved as compared to the previous 3000 RPM calculation. This is most likely due to the higher resolution of the grid in the wake. Again, the mean azimuthal velocity in the wake is predicted to be smaller by the simulations as compared to the PIV results.

In Figure 6.16 the mean azimuthal velocity is again shown along chord $O_3 - C$ for the 1500 RPM simulation. In addition to its value on the midplane we also plot the value of the azimuthal velocity integrated across a 0.5 mm thick region centered at the midplane. Here we find that the mean velocity increases (since the velocities nearer the disks are higher), but the agreement between the simulation results and the PIV does not improve very much. We therefore conclude that a combination of factors (such as modeling assumptions, grid resolution, experimental technique) could be the reasons for this discrepancy.

6.3.2 Frequency spectra

In this Section we briefly characterize the turbulence in the flow field using frequency spectra. Since PIV measurements are generally performed at very low sampling rates experimental spectra are not available for comparison.

Figure 6.17 shows the frequency spectra of the turbulent kinetic energy at various positions along chord $O_3 - A$ for the 1500 RPM simulation. We notice that the energy content is higher at both ends of the chord, i.e. near the origin O_3, which lies in the wake of the arm, and near point A, which is at the shroud. Figure 6.18 plots an average of the spectra depicted in Figure 6.17. Plotted alongside is a line representing the -5/3rd power law. In Chapter 3 we used Taylor's hypothesis to plot a similar comparison between the energy spectrum and Kolmogorov's model for the energy cascade. In this case it suffices to directly compare the frequency spectrum of the turbulent kinetic energy (in Hertz, not wavenumbers) with the ideal power law. The average energy spectrum appears to approximate the -5/3rd power law in the mid-frequency range, indicating the presence of a turbulent energy cascade.

Figure 6.19 shows a comparison of the spectra between the low velocity and high velocity regions. At the hub we expect the flow to be less turbulent than at the outer diameter, and this is clearly demonstrated by the spectra in Figure 6.19. In both spectra there is still a mid-frequency band that approximates the -5/3rd power law, indicating the presence of a turbulent cascade.

Finally, Figure 6.20 shows the contribution to the RMS velocity from different frequency bands, along chord $O_3 - A$. Here contributions are broken up into the following bands: 0-1 kHz, 1-2 kHz, 2-6 kHz and 6-10 kHz. The figure indicates that a large portion of the contribution is from the low 0-2 kHz frequency bands; while in the regions of higher RMS there is a marginal increase in contributions from the 2-6 and 6-10 kHz bands.

A similar analysis is done for the upstream chord $O_3 - B$ (1500 RPM) in Figures 6.21 to 6.23. Again, we see a mid range frequency band that displays the power law of the inertial cascade. And finally, Figures 6.24 to 6.26 present the same data along chord $O_3 - B$ (1500 RPM). The chord $O_3 - B$ is almost entirely in the wake of the arm. However, in comparing the energy spectrum from Figure 6.25 with the ones shown previously in Figures 6.18 and 6.22 we notice that the energy spectrum falls off more rapidly in the wake than at the upstream locations. The range of frequencies for which the -5/3rd power law is valid is much smaller, and for frequencies larger than about 1kHz the spectrum decays faster than the -5/3rd line. This is arguably an indication of excessive SGS dissipation in the wake and could be the cause of the discrepancies mentioned in the previous section.

6.4   Tests of Numerical Dissipation

In most commercial codes the grid-based dissipation (i.e. the numerical dissipation) cannot be quantified easily because the software source code is hidden from the user. However, using a simple test outlined below it is possible to estimate the grid-based dissipation.

Initially we run an LES case for a sufficiently long duration of time, such that the kinetic energy and windage are converged to their mean values. At a particular time we turn off (stop) the disk rotation thereby stopping all energy input into our computational domain. Since the domain is closed (unlike the the the model in the previous Chapter) no kinetic energy may enter of exit the domain externally.

With a statistical steady rotating flow but a stationary set of disks we run three simulations for a few nominal time steps (say 50):

1. We turn off the SGS dissipation (turn off LES) and set the fluid viscosity to zero. Hence, any loss of energy is only due to the numerical dissipation

2. Next, we turn on molecular viscosity, but keep SGS dissipation turned off – in this case, any loss on energy is due to viscous and numerical dissipation

3. Next, we run a full LES with viscous and SGS dissipation – here all three sources of dissipation cause the energy to decay

Based on the kinetic energy of these three simulations the numerical, SGS and viscous dissipation may be estimated algebraically. Any of the above simulations may not be run with a rotating disk, because viscosity is the only mechanism of energy input into the domain by the rotating disks (Theoretically SGS dissipation goes to zero at the walls).

Such an exercise helps to quantify the relative contribution of each source to the dissipative processes going on in our simulations. Of course, results presented here must be interpreted with caution. Since the high speed flow in the boundary layer adjacent to the disk suddenly experiences a stationary wall a large amount of viscous dissipation is expected (again, SGS dissipation goes to zero at the wall, cubically). This is indeed the case as seen in the following paragraphs.

Figure 6.27 shows the contribution to dissipation from different sources as a function of the number of time steps using the simplified technique outlined above. Our results show that at the start of this numerical experiment (i.e. the first time step of the stationary disks simulation), numerical dissipation is approximately 18%. Viscous dissipation is the largest contributor, approximately, 58%, while SGS dissipation accounts for about 24%.

On intergating the equations in time the viscous dissipation increases (due to the smearing out of the sharpest gradients, associated with the smallest eddies) and numerical dissipation decreases. This leads us to the conclusion that the numerical dissipation is smaller for low frequency (i.e. low wavenumber modes) structures, for which viscous dissipation in the primary mode of loss of energy. The SGS dissipation also reduces, again, corroborating the fact that the turbulence intensity of the flow reduce as the calculation progresses.

In interpreting these results the observation of large viscous dissipation has been explained above. In real disk drive computations we expect the viscous dissipation to be smaller, and the SGS dissipation to be a larger contributor. Nonetheless, we have been able to quantify the numerical dissipation, which is about 18%. Continuing such calculations for longer durations is generally not appropriate since the differences between each simulations widen, and algebraical subtractions of the kinetic energy to determine the three contributions is no more valid.

6.5 Conclusions

This Chapter compares results from our simulations with the PIV experiments of Barbier (2006). It is a continuation of our experimental validation effort and seeks to provide the reader certain confidence in the results we have presented in the other Chapters.

We noticed that upstream of the arm the magnitude of mean velocities and RMS velocities also agree well with the experimental predictions. The location of the spatial features of the RMS velocities (e.g. spikes) also agree well. The cross terms of the Reynolds stress tensor also compare favorably.

Downstream, both mean radial and azimuthal velocities do not agree very well. Some possible reasons that we mention, are: a poorly resolved grid which dissipates the smaller features of the flow (as we saw in the rapidly decaying frequency spectrum), or experimental inaccuracies introduced by integrating the velocities on the thickness of the measurement plane.

A simple numerical experiment also helped us to quantify the artificial dissipation introduced the entire grid. It showed that the major contributor (58%) to dissipation is through direct viscous action (most likely at the rotating disks) while the SGS dissipation accounted for about 28% of the dissipative process.

In the last 4 Chapters the focus of the work was in demonstrating adequate sophistication in our simulations. For this reason, we devised and executed rigorous comparisons of SGS models, commercial CFD codes and validated the simulations against experiments. The next Chapters 7-9 of this dissertation are more application oriented – Chapter 7 describes the computation of disk vibrations, while Chapter 8 describes the use of air-flow mitigation devices. Chapter 9 concludes this dissertation and deliberates a few ideas for quick and accurate solutions to the TMR problem.

6.6   Tables

Table 6.1: Relevant simulation data

	Simulation 1	Simulation 2
RPM	3000	1500
Reynolds Number	9190.8	4595.4
LES Model	Algebraic Dynamic	Algebraic Dynamic
Number of cells	1,378,344	2,619,246
Average resolution	0.4848 mm	0.3804 mm
Number of Axial Cells	32 over 4.65 mm	48 over 4.65 mm
Number of Processors	8	32
Time step	1×10^{-5} s	5×10^{-5} s
Number of Revolutions	6000 steps = 3 revs	4800 steps = 6 revs
Computational Time	28.351 days	23.484 days

6.7 Figures

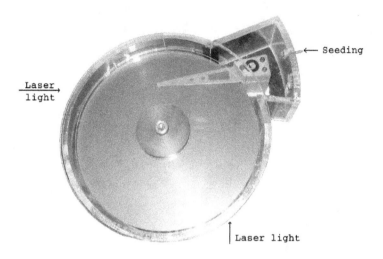

Figure 6.1: Photograph showing the experimental setup of Barbier (2006)

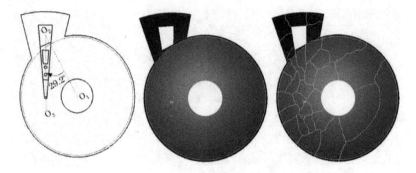

Figure 6.2: Schematic showing the experimental setup (left) the computational domain (center) and parallel, 32 processor based, domain decomposition (right).

Figure 6.3: Closeup three dimensional view showing the e-block arm, suspensions and sliders, used in the computational model

Figure 6.4: Top view of the computational grid used for 1500 RPM simulation

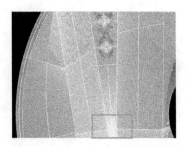

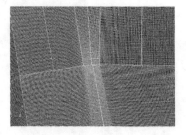

Figure 6.5: Closeup of grid, showing the increased resolution upstream
and downstream of the e-block arm /
suspensions. Highlighted rectangle is
shown in Figure 6.6

Figure 6.6: Further closeup of grid,
showing the increased resolution at the
slider location

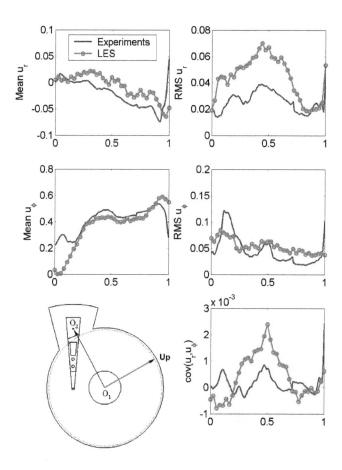

Figure 6.7: Mean and RMS of u_r and u_ϕ, and $\langle \overline{u_r}\,\overline{u_\phi} \rangle$ cross term of the Reynolds stress, shown along highlighted chord (disk speed 3000 RPM)

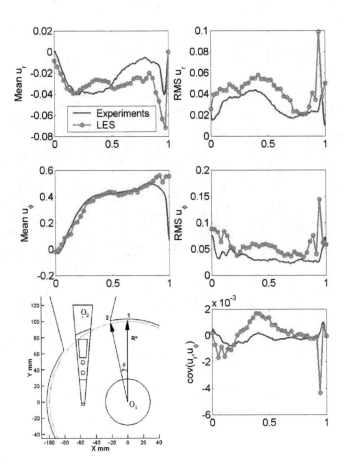

Figure 6.8: Mean and RMS of u_r and u_ϕ, and $\langle \overline{u_r}\,\overline{u_\phi} \rangle$ cross term of the Reynolds stress, shown along chord $O_1 - 1$ in the bottom left sub-figure (disk speed 3000 RPM)

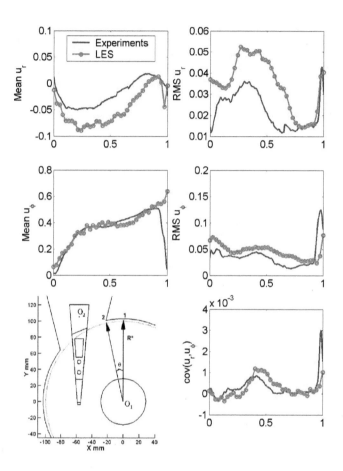

Figure 6.9: Mean and RMS of u_r and u_ϕ, and $\langle \overline{u_r}\, \overline{u_\phi} \rangle$ cross term of the Reynolds stress, shown along chord $O_1 - 2$ in the bottom left sub-figure (disk speed 3000 RPM)

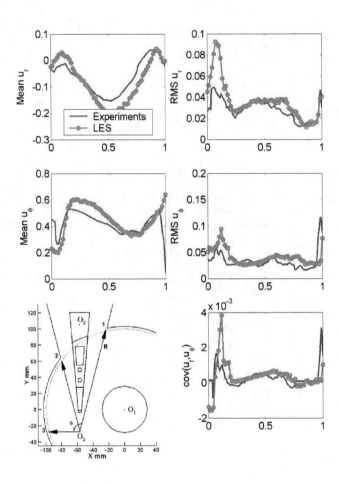

Figure 6.10: Mean and RMS of u_r and u_ϕ, and $\langle \overline{u_r}\,\overline{u_\phi} \rangle$ cross term of the Reynolds stress, shown along chord $O_3 - 1$ in the bottom left sub-figure (disk speed 3000 RPM)

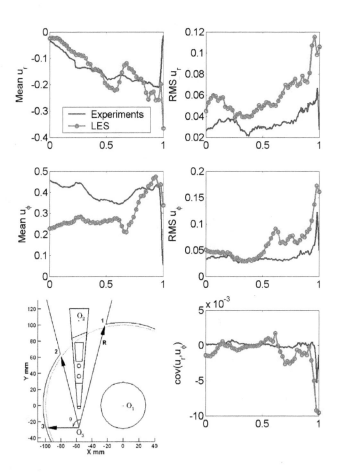

Figure 6.11: Mean and RMS of u_r and u_ϕ, and $\langle \overline{u_r}\,\overline{u_\phi} \rangle$ cross term of the Reynolds stress, shown along $O_3 - 2$ in the bottom left sub-figure (disk speed 3000 RPM)

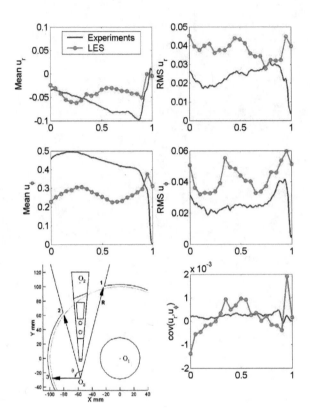

Figure 6.12: Mean and RMS of u_r and u_ϕ, and $\langle \overline{u_r}\, \overline{u_\phi} \rangle$ cross term of the Reynolds stress, shown along chord $O_3 - 3$ in the bottom left sub-figure (disk speed 3000 RPM)

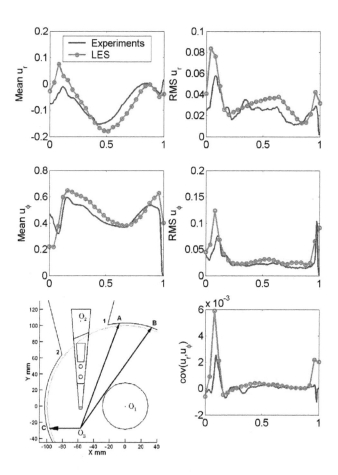

Figure 6.13: Mean and RMS of u_r and u_ϕ, and $\langle \overline{u_r} \, \overline{u_\phi} \rangle$ cross term of the Reynolds stress, shown along chord $O_3 - A$ in the bottom left sub-figure (disk speed 1500 RPM)

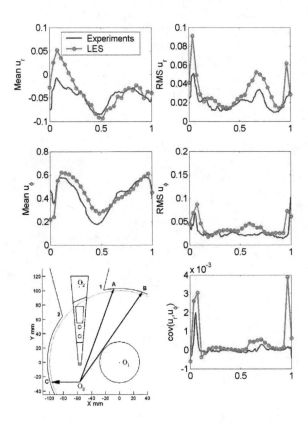

Figure 6.14: Mean and RMS of u_r and u_ϕ, and $\langle \overline{u_r}\,\overline{u_\phi} \rangle$ cross term of the Reynolds stress, shown along chord $O_3 - B$ in the bottom left sub-figure (disk speed 1500 RPM)

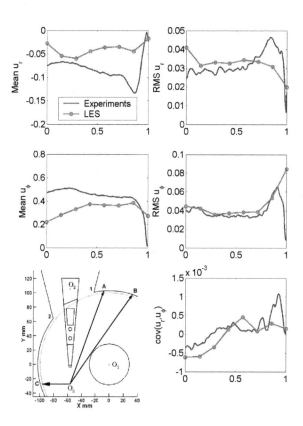

Figure 6.15: Mean and RMS of u_r and u_ϕ, and $\langle \overline{u_r}\ \overline{u_\phi} \rangle$ cross term of the Reynolds stress, shown along chord $O_3 - C$ in the bottom left sub-figure (disk speed 1500 RPM)

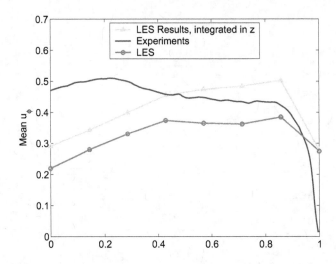

Figure 6.16: $\langle \overline{u_\phi} \rangle$ shown along chord $O_3 - C$, from Figure 6.15. Comparison is made between $\langle \overline{u_\phi} \rangle$ ad values of $\langle \overline{u_\phi} \rangle$ integrated partially in the axial direction (disk speed 1500 RPM)

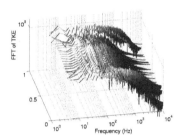

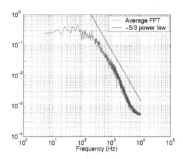

Figure 6.17: Spectrum of turbulent kinetic energy, along chord $O_3 - A$ shown in Figure 6.15 (disk speed 1500 RPM)

Figure 6.18: Spectrum of turbulent kinetic energy, averaged over the length of chord $O_3 - A$ shown in Figure 6.15 (disk speed 1500 RPM)

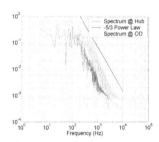

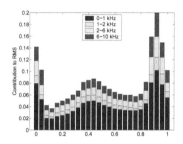

Figure 6.19: Comparison of spectra of turbulent kinetic energy, between the hub and the outer radius (disk speed 1500 RPM)

Figure 6.20: Contribution to RMS of velocity from different frequency bands, along chord $O_3 - A$ shown in Figure 6.15 (disk speed 1500 RPM)

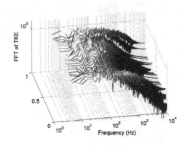

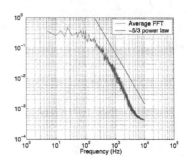

Figure 6.21: Spectrum of turbulent kinetic energy, along chord $O_3 - B$ shown
in Figure 6.15 (disk speed 1500 RPM)

Figure 6.22: Spectrum of turbulent kinetic energy, averaged over the length
of chord $O_3 - B$ shown in Figure 6.15
(disk speed 1500 RPM)

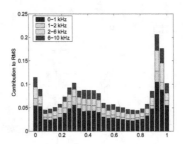

Figure 6.23: Contribution to RMS of velocity from different frequency bands, along chord
$O_3 - B$ shown in Figure 6.15 (disk speed 1500 RPM)

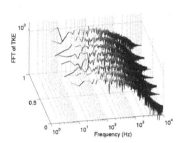

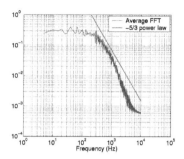

Figure 6.24: Spectrum of turbulent kinetic energy, along chord $O_3 - C$ shown
in Figure 6.15 (disk speed 1500 RPM)

Figure 6.25: Spectrum of turbulent kinetic energy, averaged over the length
of chord $O_3 - C$ shown in Figure 6.15
(disk speed 1500 RPM)

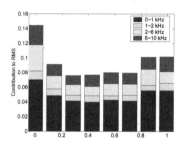

Figure 6.26: Contribution to RMS of velocity from different frequency bands, along chord
$O_3 - C$ shown in Figure 6.15 (disk speed 1500 RPM)

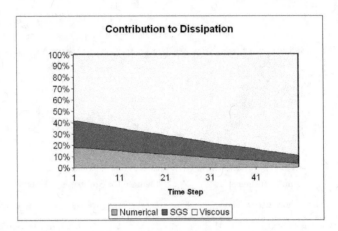

Figure 6.27: Percentage contribution to dissipation from three sources: Viscous, SGS and Numerical

Chapter 7

Computing Disk Vibrations

7.1 Introduction

In this Chapter the effect of the airflow on the rotating disk is examined. The rotating disk is the principal mechanical component of a hard disk drive, and it's dynamic stability in transverse motion is of great importance to the successful operation of the hard drive. As mentioned in Chapter 1 over the years the need for faster data transfer rates has lead to an increase in rotational speeds of disk drives. To counter the adverse effects of high speed rotation disk inertia and stiffness has generally been increasing. However, the dynamic stability of the disk itself is not the only concern. Due to the non-linear coupled nature of the head-disk interface transverse vibrations of the disk may cause significant off-track motions of the read/write head. In more extreme situations, disk vibrations may lead to the slider crashing on the disk, which results from a breakdown of the air bearing, possibly damaging the disk and/or the slider and contaminating the drive with wear particles.

Increasing spindle motor speeds have also led to the possibility of disks achieving their critical speed, which is the minimum rotational speed at which the backward traveling waves reach zero frequency and become standing waves on the disk. Due to the stationary forcing of the air bearing slider, such standing waves quickly become resonant, leading to high amplitude vibrations of the disk. Experimental data shows that current disk drives operate at approximately 40-50% of their critical speeds, but increasing drive RPMs and the need for thinner disks may bring such disks closer to the critical resonance. Disks which operate in the supercritical speed regime (e.g. floppy disks which are designed to do so) have the additional risk of undergoing hydrodynamically coupled resonant vibrations called "flutter". While this is not a concern for thicker and stiffer hard disks the air flow results presented in this Chapter may also be used as a model to simulate the aeroelastic vibrations of floppy disks.

The specific goals of this Chapter are as follows:

1. To study the flow on both sides of a single spinning disk in a fixed hard drive casing, using LES

2. To characterize the pressure loading on the disk and bring out the essential physical processes that take place in such flows

3. To develop an efficient and accurate solver to compute the free or forced vibrations of a spinning disk and to test the solver comprehensively

4. To use the solver to compute the response of the disk to realistic loadings, thus providing the disk drive research community more realistic simulation results of the flow-induced disk vibrations

7.1.1   Prior Work

The vibrations and stability of a spinning disk have been studied for almost a century. There have been significant advances both in the fluid mechanics of flows surrounding spinning disks and the structural vibrations induced by such flows. A number of authors have studied the aeroelastic stability (stability to self-excited vibrations) of spinning disks. For a summary of this research the reader is referenced to citations [1] to [12] in Kang and Raman (2004). Some investigations have used unenclosed rotating disks while others have used more realistic enclosed rotating disks. Most of the efforts, however, have used either *ad-hoc* rotating damping operators to model the surrounding flow or hydrodynamic lubrication theory to model the coupling of the flow with the disk. Generally such models based on compressible potential flow are suitable for floppy disks or circular saws, where the air film thickness is small. In hard disk drives, in addition to a large air film thickness (2-3 times the thickness of the disk), the azimuthal symmetry of the flow is broken by the presence of the actuator. This obstruction causes rapid unsteady motions in the wake its sheds, as we saw in Chapter 2. The turbulent wake is transported (and dissipated) with the rotating disks and comes around to flow over the obstruction again. It is unlikely for such a complicated turbulent flow to ever be described by an analytical model. Moreover, given the complicated and frequently changing design of actuator components (such as the suspension) the only feasible method to compute realistic disk vibrations would be to solve for the flow in a separate CFD calculation and use the resulting pressure and shear data to compute the disk vibrations.

For commercially available hard drives experimental results suggest (Fukaya et al. 2002) that the vibrations of the disk are independent of the instability of the flow. Moreover, in typical hard disk drives the experimentally measured vibrations of disks (due to all sources) are less than 0.1% of the width of the entire hard drive enclosure and the maximum linear speed of disks is approximately 10% of the acoustic speed in air. All this implies that compressible effects of the flow may be small while the turbulence induced effects on the disk vibration may be larger.

In general accurate information about disk vibrations computed using CFD based air pressure data is lacking.

Tatewaki et al. (2001) provides LES results of the a coupled CFD-structural calculation in which they report the vibrations of the disk with and without a simplified obstruction. Yuan et al. (2004) computed the flow in a deformed disk enclosed inside a casing, but with an open shroud. They present results for the pressure and shear forces on the disk in its deformed umbrella-like shape as a function of the Reynolds number. Unfortunately, their model did not have any obstruction (e.g. an actuator), and hence it resulted in a completely different flow field from what is encountered in a disk drive. Also the deflection of the disk was unreasonably large: 25% of the width of the entire hard drive enclosure, while it is typically of the order of 0.1%.

Imai (2001), Chang et al. (2002) and Guo and Chen (2001), provide experimental measurements of disk vibrations. These are then converted to off-track slider motions using the mode shapes of the disk, and they are correlated with the position error signal.

7.2 Theoretical Background

The equations of transverse motion of a spinning disk are well-known since Lamb and Southwell (1921). The governing equations may be extended to include the forcing due to the airflow excitation and the air bearing force due to the slider. These equations are given by,

$$\rho h \left(w_{,tt} + 2\Omega w_{,t\phi} + \Omega^2 w_{,\phi\phi} \right) + \frac{Eh^3}{12(1-\nu^2)} \nabla^4 w - \frac{h}{r} \left(\sigma_{rr} r w_{,r} \right)_{,r} \tag{7.1}$$
$$- \frac{h\sigma_{\phi\phi}}{r^2} w_{,\phi\phi} = \Delta p + \frac{1}{r}\delta(r-\xi)\delta(\phi)f_z$$

Here w is the transverse displacement of a point on the disk, ρ is the density of the disk material, h is the thickness of the disk, Ω is the speed of rotation of the disk, E and ν are the Young's modulus and Poisson's ratio of the disk material. The slider is modeled to exert a force f_z at a radius of ξ at $\phi = 0$ using the Dirac delta function $\delta(:)$. The unsteady distributed loading due to the airflow is given by $\Delta p = \Delta p(r, \phi, t)$. In polar coordinates, the biharmonic operator is given by,

$$\nabla^4 = \left(\frac{\partial^2}{\partial r^2} + \frac{1}{r}\frac{\partial}{\partial r} + \frac{1}{r^2}\frac{\partial^2}{\partial \phi^2} \right)^2 \tag{7.2}$$

Eqn. 7.1 assumes that the material of the disk is homogeneous, isotropic and linearly elastic (Hookean). The transverse displacements of the disk are assumed to be much smaller than the it's thickness ($w \ll h$) and the stress state of the disk is assumed to be one of generalized plane stress. Ω is also assumed to be constant. For a derivation of the full nonlinear equations of motion the reader is referred to Chung et al. (2000). The self adjoint stiffness operator,

$$\mathbb{K}[w] = \frac{Eh^3}{12(1-\nu^2)} \nabla^4 w - \frac{h}{r} \left(\sigma_{rr} r w_{,r} \right)_{,r} - \frac{h\sigma_{\phi\phi}}{r^2} w_{,\phi\phi} \tag{7.3}$$

models the bending stiffness as well as the stiffness caused by the membrane stress tensor σ. The diagonal terms of the membrane stress tensor are σ_{rr} and $\sigma_{\phi\phi}$, while the off-diagonal term $\sigma_{r\phi}$ is assumed to be zero, by which we are assuming that the membrane stress in axisymmetric. The generalized plane stress expressions for the stress components σ_{rr} and $\sigma_{\phi\phi}$, are,

$$\sigma_{rr} = c_1 + \frac{c_2}{r^2} - \frac{3+\nu}{8}r^2\Omega^2 \tag{7.4}$$

$$\sigma_{\phi\phi} = c_1 - \frac{c_2}{r^2} - \frac{1+3\nu}{8}r^2\Omega^2 \tag{7.5}$$

The constants c_1 and c_2 may be determined by the application of boundary conditions. The most commonly used model is to require that the displacement goes to zero at the inner clamp and the stresses go to zero at the free outer rim of the disk. (For some other models we refer the reader to the thesis of D'Angelo (1991)). Finally, the constants may be evaluated as,

$$c_1 = \Omega^2 \left(\frac{1+\nu}{8}\rho \frac{(\nu-1)R_o^4 - (3+\nu)R_i^4}{(\nu-1)R_o^2 - (1+\nu)R_i^2} \right) \tag{7.6}$$

$$c_2 = \Omega^2 \left(\frac{1-\nu}{8}\rho R_i^2 R_o^2 \frac{(1+\nu)R_o^2 - (3+\nu)R_i^2}{(\nu-1)R_o^2 - (1+\nu)R_i^2} \right) \tag{7.7}$$

where R_i and R_o denote the inner and outer radii of the disk.

There are four radial boundary conditions associated with Eqn. 7.1. At $r = R_i$, the transverse displacement and its rotation (slope) are zero,

$$w = 0; \qquad \frac{\partial w}{\partial r} = 0 \tag{7.8}$$

and at $r = R_o$ the radial bending moment and the shear force reaction are zero,

$$w_{,rr} + \nu \left(\frac{1}{r}w_{,r} + \frac{1}{r^2}w_{,\phi\phi} \right) = 0; \qquad (\nabla^2 w)_{,r} + (1-\nu)\frac{1}{r}\left[\left(\frac{1}{r}w_{,\phi} \right)_{,r} \right]_{,\phi} = 0 \tag{7.9}$$

Note that Eqn. 7.1 does not include any damping terms. Material damping of aluminum substrates used in hard disks is known to be significant (Kim et al. 2000). Hosaka and Crandall (1992) modeled this material damping as a term proportional to $\nabla^4 w_{,t}$. The other sources of damping are due to the clamping of the disk at its inner radius and the shear stress due to the drag of the flow on the disk surface. In our simulations each source of dissipation was not treated separately, but a global dissipation matrix was constructed from a linear combination of the mass and stiffness matrices (see Section 2.6.1). In supercritical vibrations of flexible disks certain types of damping (e.g. acoustic damping and material damping) are known to be cause the onset of flutter instabilities in which case different sources of damping need to be modeled carefully (Kang and Raman 2004). However, for our lower speed subcritical disks a single proportional damping operator suffices. The amount of dissipation for a range of frequencies was selected based on the aeroelastic parameters given in the experimental data of Kim et al. (2000). Typically the first mode of the disk is damped at about 0.02% of critical damping.

7.3 Numerical Methods

Our calculations were broadly divided into two parts. In the first part we calculated the flow of air in a simulated hard disk enclosure using CFD-ACE (Dynamic Model) and recorded the pressure data on the two sides of the disk as functions of time. In the next part of the calculation we used this time-varying pressure data to compute the vibrations of the disk using a software program developed in house. The vibrations of the disk were not fed back into CFD-ACE; thus the coupling was purely one directional, i.e. from the flow to the structure.

7.3.1 CFD methods

In our CFD work a single 3.5 inch disk enclosed inside a fixed casing was used. The model used here was based on the model used in Chapter 2, Figure ?? and was modified to simulate both sides of a single disk instead of the space between two rotating disks.

Two cases were simulated:

1. Case 1: A single e-block arm, suspension, base plate and slider was used. The slider was in contact with the disk on the bottom surface only.

2. Case 2: Two e-block arms, suspensions, base plates and sliders were used. The disk was symmetrically actuated on both of it's surfaces.

As the simulation model setup has changed when compared to the model in Chapter 2 relevant details of the model geometry are given in Table 7.1. The CFD modeling data is summarized in Table 7.2, boundary conditions are described in Table 7.3 and finally, the information regarding the grid is give in Table 7.4. A top view of the grid is also shown in Figure 7.1. With respect to Figure 7.1 we define a coordinate system, whose origin is at the center of rotation of the disks. Azimuthal angles are then defined counter-clockwise with respect to a horizontal line through the origin. Two angular positions are of special importance: the angle where the shroud separates from its circular shape to accommodate the actuator (located at about 220 degrees) and the angle where the shroud reattaches itself closely to the disk periphery (located at about 340 degrees). Figures 7.2 and 7.3 show the refinement of the grid in the vicinity of the actuator arm. Finally three dimensional outline views of the models are shown in Figures 7.4 (Case 1) and 7.5 (Case 2). Calculations were carried out on Linux based clusters, utilizing 64 CPUs at once. In spite of the massive computational power, our simulations needed to run for several weeks.

7.3.2   Structural modeling

A finite-difference code based on central differencing was developed to simulate the vibrating disk. The linearity of Eqn. 7.1 and the periodicity of the azimuthal coordinate makes it a prime candidate for a hybrid-spectral method. Firstly, the primary variable w (transverse disk vibration) is Fourier transformed in the periodic direction (ϕ) resulting in a PDE of independent variables r and t.

$$w(r,\phi,t) = \sum_{m=-N/2+1}^{N/2} \hat{w}(r,t)e^{im\phi}; \qquad \frac{\partial}{\partial\phi} \to im \qquad (7.10)$$

Transforming Eqn. 7.1, the following equation is obtained,

$$\rho h\left(\hat{w}_{,tt} + 2im\Omega\hat{w}_{,t} - m^2\Omega^2\hat{w}\right) + \frac{Eh^3}{12(1-\nu^2)}\hat{\nabla}^4\hat{w} - \frac{h}{r}\left(\sigma_{rr}r\hat{w}_{,r}\right)_{,r} \qquad (7.11)$$

$$+ \frac{m^2h\sigma_{\phi\phi}}{r^2}\hat{w} = \Delta\hat{p} + \frac{1}{r}\delta(r-\xi)\delta(\phi)\hat{f}_z$$

where,

$$\hat{\nabla}^4 = \left(\frac{\partial^2}{\partial r^2} + \frac{1}{r}\frac{\partial}{\partial r} - \frac{m^2}{r^2}\right)^2 \qquad (7.12)$$

and $\Delta\hat{p}$ and $\hat{f}_z$ are the corresponding transformed pressure differential and slider force on the disk.

Central differencing is used for the radial direction. The Laplacian operator $\nabla^2 w$ is treated specially due to the polar coordinate system,

$$\nabla^2 w \quad = \left(\frac{\partial^2}{\partial r^2} + \frac{1}{r}\frac{\partial}{\partial r} - \frac{m^2}{r^2}\right)w = \left(\frac{1}{r}\frac{\partial}{\partial r}\left(r\frac{\partial w}{\partial r}\right) - \frac{m^2 w}{r^2}\right) \qquad (7.13)$$

$$\Rightarrow \nabla^2 w_i \quad \approx \frac{1}{r_i}\frac{1}{\Delta r^2}\left(r_{i+\frac{1}{2}}(w_{i+1}-w_i) - r_{i-\frac{1}{2}}(w_i-w_{i-1})\right) - \frac{m^2 w_i}{r_i^2}$$

This method is second order consistent and is easy to implement in an implicit time integration method. The other terms of Eqn. 7.1 are discretized using standard central differences. Due to the spectral representation of one dimension of the primary variable derivatives in ϕ are exact and are computed much faster than conventional finite differences. Moreover, due to the linearity of the equations the different Fourier modes may be integrated independent of each other and forward and backward transforms need only be taken for initializing or post-processing the calculation.

7.3.3   Treatment of Boundary Conditions

The boundary conditions of Eqns. 7.8 and 7.9 are also discretized using central differencing in the same manner as the governing equation. The governing equation (Eqn. 7.1) is discretized and solved for on all radial points in the domain. Since the central-difference stencil consists of 5 points (2 on each side of the central node), the numerical method requires the solution value at 2 additional points beyond the inner and outer boundaries of the disk. The addition of these "ghost points" is helpful in maintaining second order consistency of our method (Thomas 1998). Finally, the solution values at the ghost points are expressed in terms of solutions in the actual domain by using the discretized boundary conditions.

7.3.4 Time integration

The well-known first order Newmark algorithm is used for integrating Eqn. 7.1 in time. Since the method is implicit a linear system needs to be inverted at every time step. However, since the Fourier modes are decoupled a single global matrix may be constructed for each mode and inverted only at the start of the calculation. Thereafter the calculation is progressed by the application of these stored inverted matrices. The principal advantage of using Newmark method over other conventional ODE integrators is the controllable numerical dissipation. Plots of the spectral radii of various Newmark methods are widely available (Fung 2003) and the Newmark two parameters (commonly referred to as β and γ) may be used to dissipate spurious high frequency oscillations in the solutions. Since the dissipation introduced is only numerical the frequencies of the modes do not shift, as they would when using a model for the material damping. Unless explicitly noted all of our simulations used the standard values of $\beta = 0.5$ and $\gamma = 0.25$ which are non-dissipative.

7.4 Validation of the code

Validation of codes is an essential element of the code development process. Before we used our code to produce and report results the code was put through a series of numerical tests and benchmarked against some well known data. First, some numerical tests of convergence are presented and then the code is validated against published results, such as for modal analysis and shock response. The comprehensive test results presented here should help readers in assessing the simulation numerical uncertainty of our work.

7.4.1 Tests of convergence

By a Taylor series expansion it is easy to see that our discretization scheme is second order consistent in the radial direction. In addition, the stability properties of the first order Newmark algorithm are also well known (unconditionally stable for $\beta \geq \frac{1}{2}$ and $\gamma \geq \frac{1}{4}\left(\frac{1}{2} + \beta\right)^2$) and carry over directly to our numerical scheme. For this reason we directly prove the convergence of our code based on numerical experimentation.

Firstly, while keeping the time step constant the radial mesh size is varied. The $\mathscr{L}_2$ norm of the absolute error (denoted by ϵ, based on the finest radial resolution (1.95×10^{-3})) is then computed and plotted as a function of the radial mesh size, as shown in Figure 7.8. The slope of the curve approximated by a linear curve fit is 1.87 which is close to the theoretical value of 2.0. This confirms that our central differencing scheme is convergent to second order in the radial direction.

Next, while keeping the radial mesh size constant the time step is varied. While reducing the size of the time step (Δt) computations are carried out for correspondingly longer durations and solution values are recorded at 10 well defined points in time. The $\mathscr{L}_2$ norm of the absolute error (again, denoted by ϵ) is plotted as a function of (Δt) in Figure 7.9. While the data points do not fit a linear curve perfectly the best curve fit to the data indicates an order of 0.7, which is close to the theoretical order of 1.0. Again, this proves that our implementation of the Newmark algorithm is first order convergent.

7.4.2   Modal Analysis

As a test of validation for our code we compare the modal frequencies obtained from our finite difference code with previously published theoretical and experimental data and also results from the commercial code ANSYS.

In the thesis of D'Angelo (1991) a steel disk of outer diameter 356 mm, inner diameter 106.7 mm and thickness 0.775 mm was used to make measurements. Table 7.5 compares the natural frequencies of the modes of this stationary disk computed using our finite difference code with the theoretical and experimental prediction of D'Angelo (1991). The modes are described by a pair of integers (d, c) such that d is the number of nodal diameters and c is the number of nodal circles. In our simulations the impulse response of the disk was used to extract the modal frequencies. The theoretical calculations used a spectral Galerkin method while the the experimental setup used an inductance based Tektronic Modal Analyzer. Also included in Table 7.5 are the results obtained from the commercial code ANSYS. The results essentially show that our code can predict the natural frequencies of a stationary disk to reasonable accuracy. Generally, the discrepancies between our frequencies and the others increase with the mode number. The maximum difference between our results and the results across all columns is about 8%.

7.4.3   Shock Response

Our finite-difference simulator can easily simulate the application of a shock to the disk. Shocks are typically simulated as an acceleration field applied uniformly to the entire disk in the form of a half sine wave whose amplitude is described in Gs. Figure 7.10 shows the response of a stationary 1 inch disk to a 200 G amplitude shock applied for 0.5 milliseconds (wavelength of the sine wave is 1 ms). Both structural and numerical damping were used in the simulation: the Newmark parameter β was set to 0.55 and γ to 0.3.

The exact same shock condition simulated by a commercial code (ANSYS) is shown in Figure 7.11 reproduced directly from Bhargava and Bogy (2005). The Figures 7.10 and 7.11 show a very close agreement between our finite difference results and the finite element results of ANSYS. Figures 7.12 and 7.13 show the corresponding frequency spectra. The difference in the dominant frequency (3300 Hz) is approximately 6.06% while the difference in the next frequency peak (5500 Hz) is approximately 2.7%.

7.4.4 Variation of natural frequencies with RPM

As a final test we demonstrate the ability of the code to predict the dynamics of disks under rotation. The modal analysis presented in Section 7.4.2 is for a stationary disk. It is well known that the (non-axisymmetric) disk modes split up into forward and backward traveling waves under rotation. Theoretically, in the absence of damping the frequencies increase for forward traveling waves and decrease for backward traveling waves, with a slope equal to $d\Omega$, where d refers to the number of nodal diameters of the mode. The speed of rotation at which the backward wave becomes a stationary wave is the called the critical speed. Beyond the critical RPM frequencies increase again, and the wave is called a reflected wave.

Figures 7.14-7.21 show the effect of rotation on several modes of the disk. Since the frequencies of the modes are fairly close to each other it is convenient to study the behavior of the modes separately. Generally, for a given number of nodal diameters changing the number of nodal circles changes the frequency considerably, hence modes with the same number of nodal diameters but different circles are visualized on the same figure. The frequencies presented here are for a 3.5 inch disk that was used in our CFD simulations. The modes of the stationary disk computed using ANSYS are given in Table 7.6, and they compare well with the results presented in this Section. Table 7.6 also gives the theoretical frequencies of the modes at 10,000 RPM.

As expected, Figure 7.14 shows that the axisymmetric modes (0,0) and (0,1) do not form forward and backward waves. Figure 7.15 shows the formation of $(0,1)_{F,B}$ and $(1,1)_{F,B}$ and Figure 7.16 shows the formation of $(0,2)_{F,B}$ and $(1,2)_{F,B}$. Interestingly, mode $(0,2)_B$ goes critical at about 20,000 RPM which is shown more clearly in Figure 7.17. Similarly, $(0,3)_B$ in Figure 7.18 also goes critical at about 27,000 RPM which is shown in more detail in Figure 7.17. Additionally, modes with 4 and 5 nodal diameters are shown in Figures 7.20 and 7.21. None of the remaining higher modes go critical below 30,000 RPM.

7.5 Discussion of CFD results

7.5.1 Characterization of Pressure loading

7.5.2 Mean and RMS values of the pressure loading

We begin by discussing the pressure loading on the disks. Figures 7.22 and 7.23 show the *resultant* mean pressure on the disks (averaged over 6 revolutions of the computational period) for Case 1 and Case 2 respectively. The resultant pressure is calculated as the sum of the pressure at the top and bottom surfaces of the disk – with positive pressures acting vertically out of the plane of the paper. Plots depicting the pressure may be non-dimensionalized by $\frac{1}{2}\rho U^2_{disk}(r)$, but we refrain from doing this since it would be misleading to compare such pressure coefficients directly. The mean pressure is close to zero for Case 2 in almost all parts of the disk because of the inherent (axial) symmetrical nature of the flow domain. The slight non-zero pressures at the edge of the disk (especially in the wake) are probably only numerical artifacts, occurring because the statistics may not have converged in that region.

The pressure distribution of Case 1 in Figure 7.22 shows some interesting features. Firstly, the resultant pressure is higher at the inner hub and lower at the outer hub. In Case 1 the lack of obstruction in the upper portion of the drive creates a strong radial pressure gradient. However, in the lower part of the model the actuator blocks part of the flow, which equalizes the radial pressure gradient and increases the pressure upstream of the arm, especially nearer to the hub. The resulting asymmetry causes higher resultant pressure at the hub. Closer to the outer periphery the shroud acts as a mechanism to equalize the pressure between the top and bottom parts of the drive, hence the resultant pressure load is smaller at the outer edges of the disk. The obstructing actuator creates a stagnating flow upstream, which leads to higher upstream pressures. This feature is also clearly manifested in Figure 7.22.

Figures 7.24 and 7.25 show the RMS of the resultant pressure variation on the disk surface. The RMS values are much higher at the edge of the disk (especially from the shroud separation to its reattachment), but the plots have a truncated color scale to accommodate most of the flow domain. Figures 7.24 and 7.25 show some similar features: 1) Large RMS pressure fluctuations at the periphery and smaller fluctuations at the inner hub, 2) A sharp increase in fluctuations at the shroud expansion, 3) Larger fluctuations in the wake formed behind the arm(s), and 4) Gradual reduction in the fluctuations downstream of the arm. Overall, the RMS for Case 2 is more than for Case 1 – which is a result of the turbulent flows on both sides of the disk. Finally, we also note that the presence of the arm causes a sharp break in axisymmetry, both in the mean and RMS. The loading process is thus non-uniformly distributed across several spatial scales (and time scales also, as we shall see) as characteristic of turbulent flows. Hence most analytical models cannot provide an accurate description of the loading.

7.5.3   Frequency contribution to the RMS

In addition to understanding the mean and RMS of the pressure loading it is important to understand its spectral content. From Chapters 2 and 5 we know that turbulent flows in hard drives are composed of a broad range of forcing frequencies, distributed typically from 0-10 kHz (the distribution is strongly a function of the Reynolds number).

The RMS fluctuations of the pressure broken up into contributions from various frequency bands are shown in Figures 7.26- 7.31 for Case 1 and Figures 7.32- 7.37 for Case 2. In each figure the RMS pressure fluctuations are plotted as a function of the radius, from OD to ID. The RMS contribution across different frequency bands may be easily computed using Parseval's theorem as done in Chapter 5.

Figure 7.26 shows the fluctuations in the near wake region (326 degrees from the origin). In this position we notice the high fluctuations in the outer portion of the disk which is being impacted by the eddies shed from the arm. These fluctuations are comparable to the mean value of the pressure itself. The majority contribution to the fluctuations is from the low frequency components, 0-1 and 1-2 kHz. Towards the ID fluctuations are much smaller, and are almost solely composed of the low frequency 0-1 kHz forcing. The peak in RMS is not located at the edge of the disk but at one position before the OD. As the flow progresses azimuthally the fluctuations on the disk damp out quickly. Figure 7.27 shows the fluctuations at 0 degrees. The RMS values are about half of what they were 34 degrees upstream. This rapid reduction of fluctuations appears across all frequency bands. As the flow progresses through the rest of the drive the overall fluctuations reduce through the effects of viscous and SGS dissipation. The effects of the wake appear to move from the OD to the MD in Figures 7.28 and 7.29, and the fluctuations are almost completely dissipated by 180 degrees in Figure 7.30. A common observation in Figures 7.28 through 7.30 is the relatively higher RMS at the edge of the disk. The higher RMS values are due to the turbulent flow in the shroud gap and the resulting forcing has RMS contributions from higher frequencies, up to 10 kHz. Finally, an interesting consequence of the shroud expansion is shown in Figure 7.31 at 236 degrees from the origin. The figure shows very high fluctuations at the outer edge of the disk with significant contributions from 0-1, 1-2 and 2-6 kHz frequencies. Apparently the shroud expansion causes massive flow separation and generation of turbulence with intensities that are comparable with those in the wake of the arm.

For completeness, Figures 7.32- 7.37 show the RMS pressure plots for Case 2, for the same angular positions as in Case 1. The Figures show that Case 2 has slightly higher fluctuations compared to Case 1 due to the turbulence generated on both sides of the disk. The Figures for Case 2 demonstrate many of the same qualities as discussed above for Case 1: high RMS fluctuations in the wake, high frequency contributions to fluctuations due to the flow in the shroud gap and very high RMS due to the shroud expansion.

7.5.4 Axial flow in the shroud

The RMS of the axial flow in the shroud gap (i.e. the component of the flow perpendicular to the plane of the disks) is shown in Figure 7.38 . For the complete azimuthal span (0-360) for three different axial positions: $z = 3.3$ corresponds to the top surface of the disk, $z = 2.8$ corresponds to the mid-plane of the disk and $z = 2.3$ corresponds to the bottom surface of the disk. The axial velocity is non-dimensionalized, not by the mean axial velocity but by the disk edge velocity (ΩR_o), since the former is close to zero. The RMS fluctuations of pressure at the same location as Figure 7.38 are shown in Figure 7.39. The strong correlation between the axial velocity fluctuations and the pressure fluctuations is clearly evident by comparing Figures 7.38 and 7.39. Finally, a three dimensional view which shows the locations of the different peaks in the RMS axial velocity fluctuations is shown in Figure 7.40.

We notice that the fluctuations are fairly constant in the shrouded portion of the drive, with a very gradual decrease in the amplitude downstream of the wake. The RMS of the fluctuations is a measure of the amplitude of the waves that the shroud gap can support. In both cases the RMS amplitude is small; close to 2% of the disk edge velocity. The presence or absence of an actuator simply does not matter as there is very little difference between the Cases 1 and 2. In agreement with our previous discussion the shroud expansion causes very high fluctuations in the axial velocity (seen as the first large spike at about 220 degrees in Figure 7.38). The axial velocity fluctuates rapidly as the flow approaches and flows over the arm. The wake is characterized by a single peak immediately downstream of the arm and several other peaks as the flow gets entrained back into the shrouded portion of the gap.

Figure 7.40 shows the location of these peaks three dimensionally. Also in Figure 7.40, at a particular axial location, three curves are drawn spanning the width of the shroud gap. For both cases one notices that the outermost curve does not show a peak of the same magnitude as the inner two curves at the shroud expansion. This indicates that the high fluctuations associated with the separation are present close to the disk edge only and apparently die down shortly beyond the disk edge.

7.5.5 Velocity profiles

Much about the flow can be understood by studying the average velocity profiles between the disk and the stationary casing walls. In Chapter 2 inter-disk velocity profiles were studied in Section 2.8.2. Figures 7.41 and 7.42 show the time averaged non-dimensional velocity profile $\left(\frac{\langle u_\phi \rangle (r)}{\Omega r} \right)$ as a function of the axial position between the disks. The central solid band in the figure represents the disk. The profiles are shown horizontally for different azimuthal positions (0-320 degrees) and vertically for three radial positions, ID, MD & OD. Each horizontal box spans the non-dimensional magnitude from 0 to 1, while each vertical box spans the entire axial length of the domain, as shown: from the bottom cover to the top cover. Azimuthal positions are measured from the horizontal with respect to the disk centers, with the actuator being located at about 280 degrees at MD.

In Figure 7.41, which is for Case 1, one immediately observes the asymmetry of the velocity profiles. As expected velocity profiles below the disk are smaller in magnitude than those above the disk, which is due to the loss in momentum by flowing over the arm. At about 240-280 degrees we observe that the flow stagnates in the space below the disk – which is most likely due to the blocking effects of the actuator. Interestingly, the flow also stagnates in the space above the disk at the ID position (i.e. close to the hub). Since there is no arm present in this portion of the drive the only reason for the formation of this re-circulating flow would be the expanding section of the geometry. To clarify: once the flow has passed through the shrouded portion of the drive it encounters an expanding section (to accommodate for the actuator). This expansion causes the flow to reverse direction close to the hub where the linear velocity of the disk is the smallest. On the bottom side of the disk the actuator blocks some of the expanded section accelerating the flow slightly – which results in a higher velocity of the flow at 280-320 degrees.

Velocity profiles for Case 2 in Figure 7.42 are more symmetric than for Case 1. Since the geometry, grid and boundary conditions are perfectly mirrored across the midplane any lack of symmetry can only be attributed to insufficient data in the averaging process. In general, the velocity profiles look very similar to a turbulent Couette flow. There is a sharp fall off from the disk velocity in the boundary layer with a large core region of nearly constant velocity flow followed by a sharp fall off to the stationary wall. It is also a general observation that the velocity profile is not completely flat in the central core region. Velocities are slightly higher in the region away from the disk. This can be mainly attributed to the axial location of the actuator. Referring to Figures 7.6 and 7.7 we see that the actuator is located closer to the disk than the fixed covers, hence the flow is decelerated in this region and accelerated closer to the covers. Figure 7.42 also shows the following features of the flow: 1) The flow stagnates near the hub, upstream of the actuator arm, 2) The velocity profile is small in the wake of the actuator and becomes fuller as it progresses azimuthally, 3) From about 80 degrees to about 200 degrees the velocity profiles are very similar across the ID-MD-OD positions, and 4) Almost all velocity profiles have an inflection point near the disk

7.5.6 Global Quantities

The non-dimensional kinetic energy (k^*, see Eqn. 5.12) and windage (W^*, see Eqn. 5.13) are plotted in Figures 7.43 and 7.44 as a function of the number of disk rotations.

From Figures 7.43 and 7.44 we notice that the kinetic energy in Case 2 is approximately 10% less than Case 1, which may be accounted for by the turbulent loss caused by the addition of an extra arm. However, the windage loss (i.e. the power required to drive the disks, or the rate of energy input to our computational domain), is about 2.5% smaller for Case 2 than for Case 1. This implies that with a slightly smaller rate of energy input the Case 2 simulation saturates at a lower energy level, indicating the increased presence of dissipative processes such as vortex shedding and separation.

In reporting the kinetic energy and windage the role of the initial conditions of the domain has been minimized by starting off both the simulations from a steady $k - \epsilon$ solution. Moreover, the figures also show that the quantities achieve steady state in about 4 revolutions of the disk which gives confidence that the simulations have achieved statistical steadiness on a global level. For this reason all quantities reported previously are only for the duration spanning of 4 to 10 revolutions of the simulations.

The coefficients of drag on the actuator arm are plotted as a function of time (with the first 4 revolutions removed) in Figure 7.45, and the corresponding frequency spectra are plotted in Figure 7.46. The drag and lift coefficients are decomposed into the off-track direction ($C_{D,off}$), the on-track direction ($C_{D,on}$) and the axial (out-of-plane) direction ($C_{D,z}$) using the projected areas of the actuator in those directions and the disk edge speed $U_o = \Omega R_o$. Figures 7.47, 7.48 and 7.49 summarize the statistics of the data in Figure 7.45. Interestingly, the figures show that the mean off track drag coefficient, $\overline{C_{D,off}}$, is higher in Case 2 than in Case 1, but the trend is opposite for the on-track direction, $\overline{C_{D,on}}$. The additional e-block arm, suspension, base plate and slider in Case 2 increase the off-track projected area by only 25%, while the on-track projected area is the same. This indicates that the presence of the symmetric arms in Case 2 modifies the pressure field in the drive, such that the pressure gradient acting across the arm (in the off-track direction) is increased while the gradient in the direction of the arm (the on-track direction) is decreased.

7.6   Structural vibrations

Next the vibrations of the disk are calculated using the code previously discussed. The vibrations are initialized from rest and the pressure loading of the airflow and the slider are used as forcing functions on the right hand side. An (r, m) grid of 32×64 was used, and the resultant pressure recorded at each of the 2048 nodes was used in the calculation. The slider was modeled as a point load (delta function) superimposed on the airflow pressure distribution.

The vibrations of the disk are plotted as a function of time in Figure 7.50 for both Cases 1 and 2. In plotting the displacements of the disk, since the vibrations under the slider are most important, various points were chosen from ID to OD passing through the location of the slider. The 6 sub-figures shown in Figure 7.50 are plotted for points located at 12.5%, 31.25%, 50% (MD), 62.5 %, 81.25% and 100% (OD) of the radial span.

We notice that the vibrations have a positive bias (mean) for Case 1 which is expected, given the positive mean pressure. The mean vibrations for Case 2 are close to zero. The vibrations of Case 2 clearly display a fundamental frequency (which from Table 7.6 is attributed to mode $(0, 1)_B$). Case 1 does not clearly display this frequency and is apparently composed of higher frequency components. The other interesting observation is that the vibration results appear to be a linear function of the radius; i.e. the sub-figures of Figure 7.50 appear to be geometrically very similar except for the different scale used to plot them. This leads to the conclusion that most of the vibration energy is in the modes with zero nodal circles (c=0). Given that the pressure fluctuations of the flow are mainly in the low frequencies (0-2kHz) (recall, Figures 7.26- 7.37), such a result is expected, since the first mode with one nodal circle (c=1) is above 5 kHz, and there are 6 modes of lower frequency than 5kHz (with their corresponding forward and backward traveling frequencies. See Table 7.6).

A summary of the vibration results is presented in Figure 7.51. It shows that the disk has a mean deflection for Case 1, due to the mean pressure bias. The RMS vibrations about that mean are smaller than those for Case 2, but the resulting motions are larger for Case 1. The mean vibrations for Case 2 should theoretically be zero, indicating that the small non-zero mean is a result of unconverged statistics. The RMS vibrations for Case 2 are large, approximately 680 nm. The results presented in Figure 7.51 are in good agreement with experimental measurements of the disk vibrations given in Imai (2001), Chang et al. (2002) and Guo and Chen (2001).

Due to the large number of modes in the 0-5Hz range the frequency spectra for the vibrations cannot be easily analyzed visually. It is more useful to understand the contributions to the RMS from different frequency bands, as done earlier for the pressure fluctuations. As seen in Figure 7.52, for Case 2, approximately 95% of the vibration energy is in the 0-1 kHz range, which contains the top three modes. A small amount of energy is contained in the 1-5 kHz range, while there is virtually no contribution above that range. That is the reason why the dominant frequency of Case 2 in Figure 7.50 corresponds to the $(0, 1)$ mode. For Case 1, approximately 20% of the energy is shifted from the 0-1 kHz range to the 1-5 kHz range. Given the mean deflection of the disk about its equilibrium, energy is transferred from the $(0, 1)$ mode to higher modes (mode # 4-6) and this results in a smaller

RMS amplitude of vibration.

7.7 Conclusions

This Chapter has described the structural response of the disk itself to the flow that is generated by its rotation. The Chapter has implicitly utilized several conclusions from previous Chapters (e.g. grid resolutions from Chapter 5, turbulence model from Chapter 3 and 4). At the same time this Chapter has extended several ideas about the flow that were first introduced in Chapter 2. To summarize the main conclusions from this Chapter:

1. We have presented a methodology for computing the vibrations of the rotating disk in a hard disk drive, dealing with much more realistic air flow models than previous attempts. Given the complicated nature of the flow calculations need to be repeated for even small changes in the drive configuration, (e.g. movement of the actuator from ID to OD), which makes a comprehensive investigation of all cases very expensive; almost impossible. However, we have provided two useful aids in the solution of such problems. 1) We have characterized the pressure loading in great detail for the two most commonly occurring cases (sliders flying on one or two sides of the disk) and 2) We have described the numerical methods to accurately compute the flow and the structural solutions. Finally, we expect that the data presented here will serve as a useful tool for benchmarking future calculations or experiments

2. In terms of the flow field we found that the mean pressure loading for Case 1 is asymmetric, arising due to the asymmetry of the geometrical configuration. The RMS vibrations of Case 2 are higher than those for Case 1, and their spectral distributions are almost identical. The spectra show most of the energy is in the 0-1 kHz range and the RMS increases with increasing radius. The wake and the shroud expansion show the highest fluctuations. While the formation of the wake behind the arm is inevitable drive designers should avoid expanding diffuser-like cross-sections which act as a source of separation and generation of turbulence. We also found that the fluctuations in the shroud gap corellate very well with the pressure fluctuations – hence it may be useful to measure the axial flow in the gap (say by PIV) when measurement of pressure is difficult. Finally, we also note that the velocity profiles between the disk and the casing resemble turbulent Couette flow, and it may be used to model flows in other cases.

3. In terms of the disk vibrations we note that 1 mm thick 3.5 inch Aluminum disks may be susceptible to critical behavior at about 20,000 RPM. Our results do not take into account changes in the flow field at those speeds and the resulting damping. Nonetheless, drives that operate in the 20,000-30,000 RPM range may be subject to these resonances. At 10,000 RPM our results show that for Case 1 the disk undergoes a mean asymmetric deflection and vibrates at a smaller magnitude in the higher modes. For Case 2 there is no mean deflection, and the disk primarily vibrates in the lowest fundamental mode.

4. A direct continuation of this work (which is beyond the scope of this book) would be to compute the resulting off-track and on-track motions of the slider-head due to both the air drag and the disk motion. While this is currently unfeasible in CFD-ACE, given the separation of scales, this could be the subject of another investigation that accounts for all of the non-linear dynamics of the slider motion.

7.8 Tables

Table 7.1: Geometry data

	Case 1	Case2
Number of disks	1	←
Number of e-block arms	1	2
Number of base plates	1	2
Number of suspensions	1	2
Number of sliders	1	2
Disk thickness (mm)	1	←
Disk diameter (mm)	76.2	←
Width of shroud gap (mm)	1	←
Length of actuator (mm)	45	←
Length of e-block arm (mm)	32.5	←
Length of base plate (mm)	6.5	←
Length of suspension (mm)	11.1	←
Thickness of e-block arm (mm)	0.8	←
Thickness of base plate (mm)	0.3	←
Thickness of suspension (mm)	0.1	←
Dimensions of slider (mm)	$1 \times 0.8 \times 0.3$	←
Number of weight saving holes in e-block arm	2	←

]

Table 7.2: CFD modeling information

Governing equations	Filtered Navier Stokes equations
Solution algorithm	SIMPLEC (Van doormaal and Raithby 1984)
Large eddy simulation model	Algebraic dynamic (Germano et al. 1991)
Type of LES filter	Top-hat (variable width)
Temporal differencing scheme	Crank Nicholson (second order)
Spatial differencing scheme (convective term)	Central differencing
Time step (seconds)	1.0×10^{-5}
Number of time steps	4800
Corresponding number of disk rotations	8
Initial conditions	Steady k-ϵ solution

Table 7.3: Boundary conditions

Disks	Rigid rotating walls, no slip
Casing	Rigid wall, no slip
Hub/base of e-block arm	Fixed (similar to a cantilever)
Slider-disk interface	Slider slips on disk
	No cells between slider and disk
All structural interfaces	Rigidly joined
(e.g. suspension+slider,	(i.e. no dimple)
e-block arm+base plate)	
All fluid-structure surfaces	walls, no slip

Table 7.4: Grid information

	Case 1	Case 2
Type of mesh	Structured grid mixed with	←
	quad-dominant unstructured cells	
Number of cells.	4,515,444	4,444,274
Avg. cell vol. (mm^3)	5.7338×10^{-3}	5.7994×10^{-3}
Avg. grid res. (mm)	0.1789	0.1796

Table 7.5: Comparison of natural frequencies of a stationary disk

Mode	Current Work (Hz)	Experiments (Hz)	% Diff	Theory (Hz)	% Diff	ANSYS (Hz)	% Diff
(0,1)	36	37.19	3.20	39.08	7.88	38.06	5.41
(0,0)	40	38.40	-4.17	39.73	-0.68	38.68	-3.42
(0,2)	46	47.10	2.34	47.46	3.08	46.20	0.44
(0,3)	77	79.78	3.48	79.18	2.75	77.10	0.13
(0,4)	124	133.08	6.82	131.64	5.80	128.30	3.35
(0,5)	189	202.18	6.52	200.16	5.58	195.27	3.21
(1,0)	236	250.18	5.67	254.18	7.15	247.31	4.57
(1,1)	246	262.38	6.24	266.23	7.60	259.00	5.02
(0,6)	269	285.71	5.85	282.68	4.84	276.19	2.60
(1,2)	280	304.79	8.13	303.88	7.86	295.50	5.25
(1,3)	348	374.92	7.18	370.12	5.98	359.53	3.21

Note: Experimental and theoretical data is from D'Angelo (1991)

Table 7.6: Natural Frequencies and modes of the disk, obtained from ANSYS

Mode No.	Frequency (Hz)	Mode (d,c)	Forward Mode @ 10K RPM	Backward Mode @ 10K RPM
1	781.56	(1,0)	948.23	614.89
2	802.71	(0,0)	–	–
3	946.65	(2,0)	1279.98	613.32
4	1610.9	(3,0)	2110.90	1110.90
5	2704.4	(4,0)	3371.07	2037.73
6	4126	(5,0)	4959.33	3292.67
7	5104.3	(0,1)	–	–
8	5355.1	(1,1)	5521.77	5188.43
9	5837.6	(6,0)	6837.60	4837.60
10	6147.3	(2,1)	6480.63	5813.97
11	7552.3	(3,1)	8052.30	7052.30
12	7826.9	(7,0)	8993.57	6660.23
13	9588.1	(4,1)	10254.77	8921.43
14	10090	(8,0)	11423.33	8756.67
15	12194	(5,1)	13027.33	11360.67
16	12624	(9,0)	14124.00	11124.00
17	14773	(0,2)	–	–
18	15071	(1,2)	15237.67	14904.33
19	15271	(6,1)	16271.00	14271.00
20	15430	(10,0)	17096.67	13763.33
21	15991	(2,2)	16324.33	15657.67
22	17599	(3,2)	18099.00	17099.00
23	18507	(11,0)	20340.33	16673.67
24	18736	(7,1)	19902.67	17569.33
25	19951	(4,2)	20617.67	19284.33

7.9 Figures

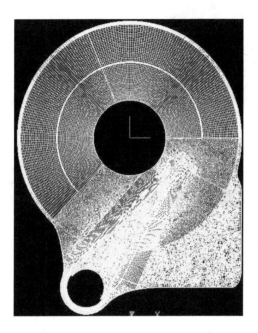

Figure 7.1: Top view of the computational grid

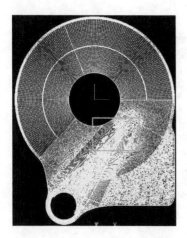

Figure 7.2: Location of closeup in Figure 7.3

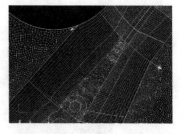

Figure 7.3: A closeup view of the refined grid upstream and downstream of the actuator

Figure 7.4: Three-dimensional simplified view of the simulated Case 1. Shown are the disk and the single arm actuating the lower surface

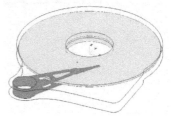

Figure 7.5: Three-dimensional simplified view of the simulated Case 2. Shown are the disk and the arms on both the disk surfaces

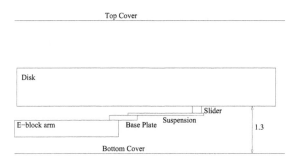

Figure 7.6: Schematic diagram showing cross section of simulation domain for Case 1

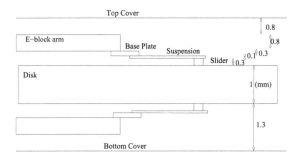

Figure 7.7: Schematic diagram showing cross section of simulation domain for Case 2

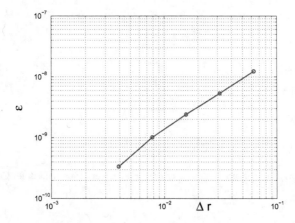

Figure 7.8: Radial convergence: $\mathscr{L}_2$ norm of the error (ϵ) as a function of the radial mesh size (Δr)

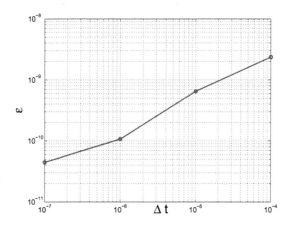

Figure 7.9: Radial convergence: $\mathscr{L}_2$ norm of the error (ϵ) as a function of the time step (Δt)

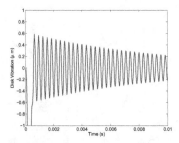

Figure 7.10: Response of a disk to 200
G shock of 0.5 milliseconds. computed
using current code

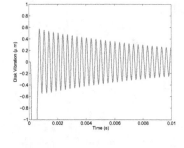

Figure 7.11:   Response of a disk to
200 G shock of 0.5 milliseconds. computed using ANSYS (Bhargava and
Bogy 2005)

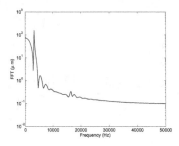

Figure 7.12: FFT of disk response to
200 G shock of 0.5 milliseconds. computed using current code

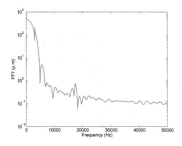

Figure 7.13: FFT of disk response to
200 G shock of 0.5 milliseconds. computed using ANSYS (Bhargava and
Bogy 2005)

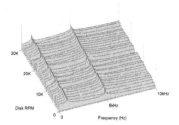

Figure 7.14: Waterfall plot showing the natural frequencies as a function of RPM. Modes (0,0) and (1,0)

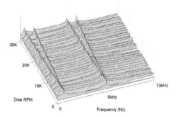

Figure 7.15: Waterfall plot showing the natural frequencies as a function of RPM. Modes (0,1) and (1,1)

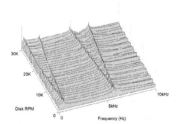

Figure 7.16: Waterfall plot showing the natural frequencies as a function of RPM. Modes (0,2) and (1,2)

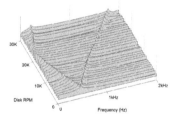

Figure 7.17: Closeup showing the critical behavior of mode $(0,2)_R$ at 20,000 RPM

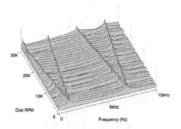

Figure 7.18: Waterfall plot showing the natural frequencies as a function of RPM. Modes (0,3) and (1,3)

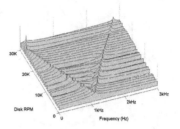

Figure 7.19: Closeup showing the critical behavior of mode $(0,3)_R$ at 27,000 RPM

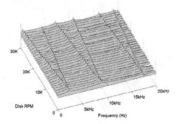

Figure 7.20: Waterfall plot showing the natural frequencies as a function of RPM. Modes (0,4), (1,4) and (2,4)

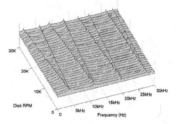

Figure 7.21: Waterfall plot showing the natural frequencies as a function of RPM. Modes (0,5), (1,5), (2,5) and (3,5)

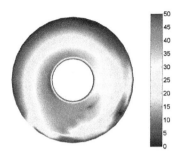

Figure 7.22: Mean resultant pressure distribution on the disk, for Case 1

Figure 7.23: Mean resultant pressure distribution on the disk, for Case 2

Figure 7.24: RMS resultant pressure distribution on the disk, for Case 1

Figure 7.25: RMS resultant pressure distribution on the disk, for Case 2

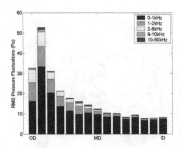

Figure 7.26: RMS Pressure fluctuations, broken down into contributions from frequency ranges. For Case 1 at 326 degrees.

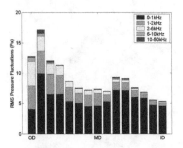

Figure 7.27: RMS Pressure fluctuations, for Case 1 at 0 degrees.

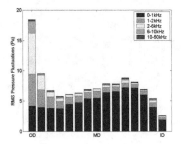

Figure 7.28: RMS Pressure fluctuations, for Case 1 at 56 degrees.

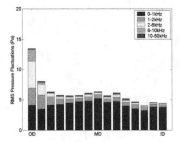

Figure 7.29: RMS Pressure fluctuations, for Case 1 at 112 degrees.

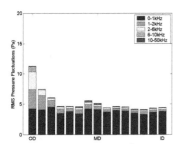

Figure 7.30: RMS Pressure fluctuations, for Case 1 at 180 degrees.

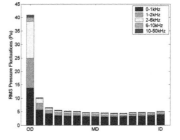

Figure 7.31: RMS Pressure fluctuations, for Case 1 at 236 degrees.

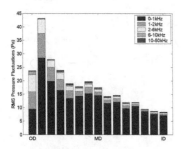

Figure 7.32: RMS Pressure fluctuations, broken down into contributions from frequency ranges. For Case 2 at 326 degrees.

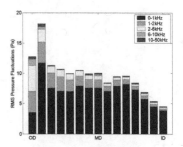

Figure 7.33: RMS Pressure fluctuations, for Case 2 at 0 degrees.

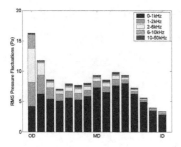

Figure 7.34: RMS Pressure fluctuations, for Case 2 at 56 degrees.

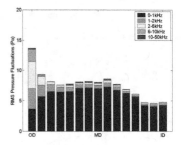

Figure 7.35: RMS Pressure fluctuations, for Case 2 at 112 degrees.

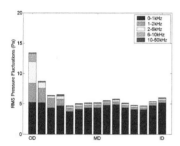

Figure 7.36: RMS Pressure fluctuations, for Case 2 at 180 degrees.

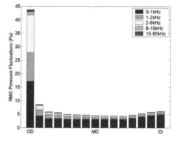

Figure 7.37: RMS Pressure fluctuations, for Case 2 at 236 degrees.

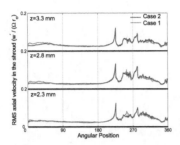 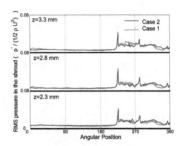

Figure 7.38: The RMS of the axial velocity at the center of the shroud gap, plotted for three different positions

Figure 7.39: The RMS of the pressure fluctuations at the center of the shroud gap, plotted for three different positions

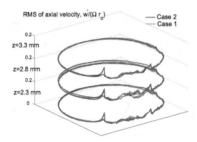

Figure 7.40: Three dimensional view showing the angular location of axial velocity fluctuations

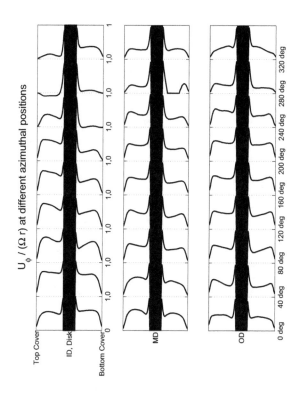

Figure 7.41: Interdisk non-dimensional velocity profile for Case 1, shown as a function of the azimuthal angle, for three positions, ID, MD and OD

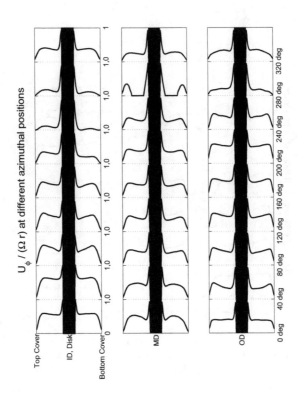

Figure 7.42: Interdisk non-dimensional velocity profiles for Case 2, shown as a function of the azimuthal angle, for three positions, ID, MD and OD

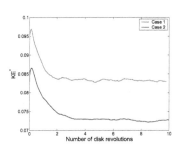

Figure 7.43: Non-dimensional kinetic energy

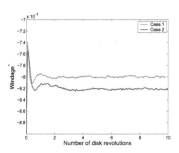

Figure 7.44: Non-dimensional windage

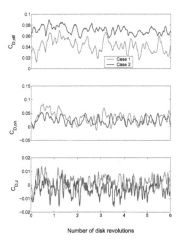

Figure 7.45: Time history of variation of coefficients of drag

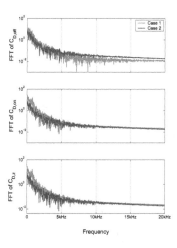

Figure 7.46: FFT of drag coefficients

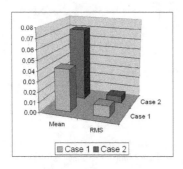

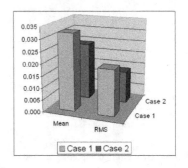

Figure 7.47: Summary of statistics for $C_{D,off}$

Figure 7.48: Summary of statistics for $C_{D,on}$

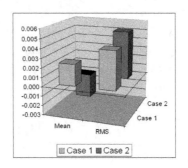

Figure 7.49: Summary of statistics for $C_{D,z}$

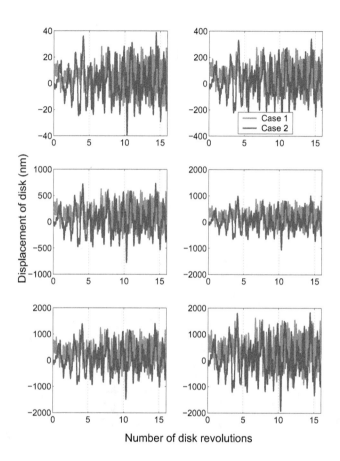

Figure 7.50: Disk vibration results showing the displacement of the disk as a function of time. Results are shown for points located at 12.5%, 31.25%, 50% (MD), 62.5 %, 81.25% and 100% (OD)

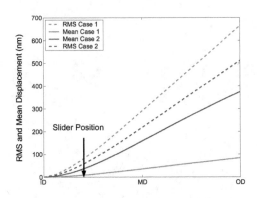

Figure 7.51: Summary of disk displacement results

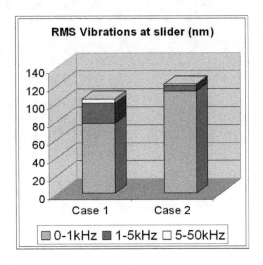

Figure 7.52: Contribution to the RMS disk vibrations under the slider from various frequency ranges

Chapter 8

The Effect of Flow Mitigation Devices

8.1 Introduction

In this Chapter we study the effect of several "flow mitigation devices" (such as spoilers, ribs and plates) that are commonly used in disk drive casings. Such modifications (several of which are listed in the patent literature) have been designed to reduce the effect of the flow on the TMR. This work does not seek to design such devices that change the flow and the resulting vibrations. Design influences such as costs, materials and manufacturability are beyond the scope of this book, but they nevertheless play an important role in the design process. In this light, the modifications studied here are some of those in current use in disk drives, and they have been selected for investigation after examining several disk drives available in the market, in late 2004. The modifications investigated are: M1: a blocking plate situated between the disks, M2: a spoiler (or deflector) located behind (downstream of) the actuator arm and M3: a similar deflector upstream of the arm. A comparison is made between the modifications M1-3 and the original disk drive without any modification, which we denote as M0.

An interesting finding from the patent literature is the effect of reverse spinning disks. Our simulations show that simply reversing the direction of rotation of the disks could reduce the vibrations of the arm and disk significantly. This is discussed in Section 8.4

8.1.1 Model Setup

The original model (M0) without any modifications simulated in this Chapter is the same as the one used in Chapter 2, Figure 2.2. A top view of this model (and the other geometrical models simulated) are shown as follows, M0 in Figure 8.1, M1 in Figure 8.2, M2 in Figure 8.3 and M3 in Figure 8.4. The relevant geometrical modifications have been highlighted for clarity. Geometrical data that is common to all simulations is given in Table 8.1, while geometrical data specific to each simulation is given in Tables 8.2-8.4. Numerical modeling information that is common to all simulations is given in Table 8.5 and boundary conditions are outlined in Table 8.6. Finally, information about the mesh used in the simulations is given in Table 8.7. The grid used for M0 was the same as that used in Chapter 2. The grids used for M1-3 were modified accordingly to account for the flow mitigation devices. The grids were qualitatively very similar and were each generated by extruding a 2-D grid in the axial direction.

In Table 8.7 the average cell volume is calculated by taking a mean of all the computational volumes in the simulation domain. The average grid resolution is the cube root of the average cell volume, which forms a measure of the representative grid size.

8.2   Flow Physics

We start by discussing some physical features of the flow and subsequently describe the more quantitative results.

8.2.1   Major Flow Features

As we saw in Chapters 2-6, flows in disk drive enclosures are highly unsteady with partly laminar and partly turbulent regions. Snapshots of the turbulent flow in our simulations are shown in the Figures 8.5 - 8.8. Plotted therein is the axial component of velocity on the midplane between the disks. Instead of choosing a monotonic scale for plotting this component of velocity we used a staggered scale (similar to an interference pattern) is used. This helps in visualizing sharp velocity gradients that characterize the turbulent eddies, which may not appear in a monotonic scale. However quantitative information about the velocity magnitude is lost in this presentation mode. Nonetheless, this is acceptable for now, since we refer to quantitative data in later sections.

As the air flows over the structures forming an obstruction it undergoes separation causing the formation of vortical structures (see (1) in Figure 8.5). The vortex shedding causes changes in the circulation around the arm, which causes fluctuation in the drag it experiences. Turbulent eddies formed in the wake of the arm are convected by the mean flow due to the disk rotation and dissipate by the time they reach an angular position of approximately 225° [1] (see (2) in Figure 8.5). The turbulence intensity of the flow coming towards the actuator arm is between 15-20% (as later explained by Figure 8.18). At the curved wall which forms the shroud one observes the presence of one or two toroidal vortices (see (3) in Figure 8.5). These structures are Göertler-type vortices formed due to three dimensional

---

[1] In describing radial and angular locations of our geometry, the origin is taken at the center of rotation of the disks (as in previous chapters).

instability of the laminar boundary layer as it flows over the concave boundary. Finally, in the region upstream of the actuator arm, where the enclosure expands to accommodate the arm, one observes separation of the flow and the formation of a turbulent region. (see (4) in Figure 8.5)

When compared with M0, M1 shows significant changes in the flow field, which is plotted in Figure 8.6. The presence of the blocking plate essentially blocks out a significant portion of the flow, forcing the rest around it. The mean velocity of the flow is reduced because the blocking plate acts in regions where the linear velocity of the disk is higher. However, vorticity shedding at the trailing edge of the blocking plate increases the turbulence of the flow approaching the e-block arm (see (5) in Figure 8.6). Also, one observes the presence of a region of flow reversal (and stagnation) near the hub. The presence of the blocking plate causes an adverse pressure gradient in the air flowing towards it, causing some portions of the flow to stagnate and reverse direction (see (6) in Figure 8.6).

In simulation M2, plotted in Figure 8.7, the presence of a thick (1.6 mm compared to a disk-to-disk spacing of 2.2 mm) downstream rib blocks a significant portion of the flow. At the midplane the mean azimuthal velocity is decreased almost everywhere in the drive. The presence of the rib causes the flow to stagnate and reverse direction in a significant portion of the drive (see (7) in Figure 8.7). The rib is a source of vorticity shedding also, which increases the turbulence intensity of the downstream flow (see (8) in Figure 8.7).

In simulation M3, which is plotted in Figure 8.8, the flow field is similar to M0, except that the upstream spoiler acts as another source of generation of turbulent eddies. The eddies shed from the top and bottom edges cause added velocity fluctuations in the upstream portion of the flow field (see (9) in Figure 8.8). This significantly increases the velocity fluctuations near base of the e-block arm and the suspension.

8.2.2   Evolution of flow

It is expected that the airflow velocity magnitude is smallest in the wake, and the flow gains momentum from the rotating disks as it flows around. To gain more insight into this process we plot the inter-disk velocity profiles at 4 points in the drive. In polar $(r \text{ (mm)}, \phi)$ coordinates, these 4 points are $(14.96, 340°)$, $(14.96, 45°)$, $(14.96, 135°)$, $(14.96, 225°)$. $r = 14.96\text{mm}$ corresponds to $1/3^{\text{rd}}$ the radial span of the disks, chosen so as to not lie within the blocking plate. The angular positions were chosen so as to not lie in the path of any modification. Data plotted in each figure is the average velocity profile over 6 revolutions of the disks. In this Chapter when a direct comparison between simulations is permitted the results are plotted using a common convention. This convention is explained in the legend given in Table 8.8.

In Figures 8.9 - 8.12 the azimuthal velocity of the flow is plotted as a function of the axial z coordinate for the above mentioned 4 points. $z = 0$ refers to the top of the bottom disk, while $z = 2.2\text{mm}$ refers to the bottom of the top disk. All figures are plotted to the same scale for convenience.

It is observed that at $340°$ in the wake M0 shows the *fullest* profile, implying that the unmodified flow is the fastest in the wake. The velocity profile for M1 is less *full* because of the presence of the blocking plate, while M2 shows a mid span flow reversal in the wake. The presence of the downstream rib and its corresponding pressure gradient causes the flow to reverse directions in the wake. Part of the flow closer to the disks flows in the direction of rotation, while the bulk of the center section flows in the reverse direction.

As the flow moves around to 45° the velocity profiles for all the simulations become *fuller* due to the diffusion of momentum from the rotating disks. M0, which is the flow without any obstructions, shows the largest magnitude, while M2 shows the smallest profile. None of the profiles show flow reversal. The width of the (laminar) boundary layer is approximately the same in each simulation. M2 shows the largest velocity gradient in the boundary layer.

At 135° M0 again shows the fullest profile, and M2 begins to show flow reversal, which is due to the presence of the downstream spoiler, approximately 180° upstream. The profile for M3 is similar to that for M0, reduced in magnitude by approximately 50%. This is a direct consequence of the upstream spoiler.

Finally at 225° the profiles for M0 and M1 are almost identical, M3 is reduced from M0 by approximately 50%, while M2 shows flow reversal. This confirms that the presence of the downstream spoiler causes a significant portion of the flow in the drive to reverse direction, mostly in the regions close to the hub.

As an aside, it is interesting to note that almost all of the *mean* velocity profiles plotted in Figures 8.9 - 8.12 satisfy Fjørtoft's criteria for instability, which asserts that a necessary condition for instability of inviscid parallel flows is that $U_{yy}(U - U_I) < 0$ somewhere in the flow, where U_I is the velocity at the point of inflection of the profile. (For Fjørtoft's Theorem see Kundu (1990)). This indicates that the mean flow profile in a disk drive enclosure (with or without the modifications) does satisfy the necessary condition for being linearly unstable in the inviscid limit. The only profiles that are stable are M0 and M3, both at 135°.

Figures 8.13-8.16 show the radial velocity as a function of the axial coordinate for the same 4 points as in Figures 8.9 - 8.12. In each figure one observes a positive spike in radial velocity immediately adjacent to the disks, as expected, due to the centrifugal effect.

At 340° in the wake, the radial velocity profiles are not too different from each other. They are mainly affected by the constraining geometry of the model, which tends to squeeze the flow in the radial space between the hub and the shroud. For this reason two peaks in the radial inflow velocity (i.e. negative radial velocity) are observed for each profile.

At 45° the radial velocity of M1 is strongly negative. This is because the blocking plate tends to bend the streamlines towards the hub. The other profiles show radial outflow, with M2 showing the largest variation across the inter-disk spacing.

At 135° M1 again shows the effect of the blocking plate, while M3 shows the effect of the upstream spoiler, both of which tend to create radial inflows.

Finally, at 225°, the presence of the upstream spoiler is clearly evident as indicated by the strong negative radial velocity profile for M3. On the other hand M1 now shows larger positive radial velocity, since beyond the trailing edge of the blocking plate lies an expansion region where the flow can radially spread before approaching the actuator.

8.2.3 Turbulence intensity

Figure 8.17 shows a schematic diagram of the disk drive enclosure with all three modifications super-imposed. Also shown in this figure are 4 chords running from the inner radius to the outer radius at angular positions of 340°: Chord 1; 45°: Chord 2; 135°: Chord 3 and 225°: Chord 4. Plotted in Figures 8.18 - 8.21 are the turbulence intensity (TI) profiles along these chords. The chord length is non-dimensionalized by the radial span of the disks. In general one observes that the TI is higher in regions closer to the hub than in regions near the outer radius. This is because, near the hub, the disk velocities are small, the flow tends to reverse direction and hence the RMS fluctuations appear to be a larger fraction of the mean.

For chord 1 in Figure 8.18 one can clearly observe a single peak in TI due to the wake of the actuator arm for M0, M1 and M3. TI values are smaller near the outer and inner radii, hence it appears that a large part of the wake fluctuation is located near midway between the outer and inner radii. Compared to M0, M1 shows significantly reduced turbulence intensity. For M2 one observes two peaks, which is due to the vortex shedding occurring from the top and bottom edges of the downstream spoiler.

M1 shows higher fluctuations in Figures 8.19- 8.21 in regions adjacent to the hub. (In Figure 8.20 the TI profile for M1 is incomplete due to the blocking plate).

M0 and M3 show remarkably similar TI profiles along each chord, indicating the presence of the upstream spoiler does not change the turbulent fluctuations along the chords being considered.

8.3 Velocities and Pressure in the vicinity of the actuator

We now shift our attention from examining the entire flow domain to examining the region close to the actuator arm. The following results pertain to velocity and pressure data at a few specific points (ranging from 1-32), which are shown in Figure 8.22. These points lie close to the face of the actuator at an axial position which is at the center of the solid structure. Points 4-10 are along the centerline of the e-block arm, while 12-22 are along the centerline of the lower suspension.

8.3.1 Velocity fluctuations

To begin, we examine the RMS of the in-plane (i.e. in the plane of the disks) velocity fluctuations. This is plotted in Figure 8.23. RMS fluctuations for M0-M3 have been plotted on separate figures for clarity.

The figure for M0 shows two distinct peaks near points 5-8. These are the fluctuations arising due to the expansion of the shroud just upstream of the e-block arm (See (4) in Figure 8.5). Two more peaks in fluctuation are observed: at point 18, due to the vorticity shedding at the slider and at points 21-22, due to the vorticity shedding from the corner of the base plate.

Comparing this to M1 it appears that M1 is able to dampen the fluctuations near the slider, but the fluctuations near the e-block arm actually increase. This is indeed a favorable effect since fluctuations near the e-block arm contribute less to actuator vibrations than fluctuations near the slider. The added fluctuations near the e-block arm are due to the vorticity shedding from the edge of the base plate (See (5) in Figure 8.6).

M2 displays less fluctuations near the base of the actuator but increased fluctuations near the region of the slider. Finally, M3 shows much higher fluctuations at the base of the actuator (points 1-5 and 29-32) and the base-plate and suspension region (points 10-15) due to the shedding of vortices from the upstream spoiler.

Further insight into the RMS fluctuations can be gained from the frequency spectra of the in-plane velocity at each point. This is plotted for M0-M3 in Figures 8.24 - 8.27. The coloration of each figure corresponds to dB amplitude of the spectrum.

Comparing Figure 8.25 to Figure 8.24 one readily observes that the blocking plate dampens the power in the spectrum at all locations except the base of the e-block arm. However, the spectra do not show significant changes near the suspension using any other modification. In fact, from Figure 8.27, it is evident that the presence of the upstream spoiler actually increases the fluctuations surrounding the actuator, especially near the base of the actuator and the leading edge of the suspension.

Plotted in Figure 8.28 are the RMS of the out-of-plane (axial) velocity fluctuations. The plot for M0 shows two significant peaks – one corresponding to the fluctuations arising from the expansion of the shroud, and the other corresponding to the vorticity shedding off the slider edge. The trailing edge of the e-block arm (region 24-30) also shows higher axial fluctuations.

In the same figure, one observes that the out-of-plane fluctuations near the slider are reduced by the presence of the blocking plate, they are favorably reduced almost everywhere in M2, but are significantly increased in M3. The upstream spoiler contributes to the significantly high out-of-plane fluctuations near the base of the e-block arm (region 2-5 and 29-32).

Plotted in Figures 8.29 - 8.32 are the corresponding frequency spectra, which provide more quantitative information regarding the out-of-plane velocity fluctuations. Again the spectrum for M1 in Figure 8.30 contains significantly less power than the spectrum for M0. A common observation from these figures is that, when a modification is used to reduce RMS fluctuations of velocity, higher frequency bands, corresponding to smaller eddies, are damped out. This implies that the energy content of the smaller eddies is reduced by the use of modifications like the blocking plate, while the energy content of the larger eddies, which is determined by the disk spacing and disk speed of rotation, remain relatively unchanged.

8.3.2   Pressure difference across the actuator

From Chapter 2 we known that form drag due to pressure produces forces 2 orders higher in magnitude than skin friction (viscous) drag. Hence we examine the RMS of pressure fluctuations along the length of the actuator. Fluctuations in pressure at the leading or trailing face of the actuator contribute to it's in-plane motions, while the less important out-of-plane pressure fluctuations acting on the top and bottom surfaces of the actuator cause bending in the suspension and e-block arm. We report only the in-plane pressure fluctuations.

Figure 8.33 is a plot of the RMS of pressure fluctuation for points 1-32. M0 shows two peaks in the RMS pressure fluctuation, the first is due to the flow separation due to the shroud expansion, while the next is due to the vorticity shedding from the slider. M1 is effective in reducing the pressure fluctuations due to the vorticity shedding from the corner of the base plate. M2 shows much smaller fluctuations near the base of the arm, but the fluctuations are increased near the suspension and base plates. No clear peaks in RMS are observed. Finally, M3 shows significantly larger fluctuations at the base of the actuator and at the location where the turbulent eddies shed from the upstream spoiler impinge on the suspension.

Figures 8.34 - 8.37 show the frequency spectra of the pressure fluctuations for M0-3. When compared to M0, M1 shows reduced frequency content in the higher frequency bands, indicating that smaller eddies (i.e. eddies of higher frequencies) contribute less to the pressure fluctuations. This is especially important in the region of the suspension (between 14-22). Figure 8.36 shows that with the addition of a downstream spoiler the frequency content of the spectrum is relatively unchanged, except that the amplitude of the spectrum is overall reduced. This suggests that although the amount of energy in pressure fluctuations has been reduced the distribution of energy over spatial scales of motion has remained unchanged. Finally, Figure 8.37 confirms that the upstream spoiler is ineffective in reducing pressure fluctuations.

8.3.3 Windage

It is expected that the cost of operation of these modifications (primarily the power required by the motor, i.e. windage) should not be prohibitively high. Plotted in Figure 8.38 is the time history of the windage calculated as a function of the disk revolutions. (To give the reader a sense of the actual power increase the dimensional windage, W, from Eqn. 2.32, in watts is plotted). One observes that although the initial conditions were inaccurate in predicting the windage, it asymptotes to a constant value in approximately 2 disk revolutions.

One also observes that M1, due to its large blocking plate, consumes the most power, while the windage loss for M2 is also high, given the flow reversal near the hub. This is expected given that the axial velocity gradients are considerably higher for M1 and M2 compared to M0 and M3 leading to higher shear stresses on the disks. The windage losses for M0 and M3 are almost identical.

8.4 Effect of Reverse Spinning disks

In this Section we examine the effect of the disk spinning in the reverse direction on the forces affecting the arm and disk. Conventionally, disk drives spin counter clockwise when viewed from top. The original idea behind reverse (clockwise) rotating disks was published by Zeng and Hirano (2005) and the authors hold a patent for the same.

In their experimental publication (Zeng and Hirano 2005) the authors claim that the flow induced vibrations of the actuator arm may be reduced by as much as 50% by spinning the disk in reverse. They also mention that disk vibration is reduced by 30%. To verify this claim and to provide computational validation of their idea we compute the flow with reverse spinning disks.

The model used in this Section differs from the four models used in the previous Sections of this Chapter. M0-3 are all models that compute the flow between two co-rotating disk. Pressure forces acting at the top and bottom of a single disk are not computed. Hence, with the intention of computing the vibrations of the disk, the model used in this Section is the same as that used in the Chapter 7, Case 2. A simplified version of the model is shown in Figure 7.5 while a schematic diagram showing the cross section of the simulation domain is shown in Figure 7.7

8.4.1   Flow features

Reversing the direction of rotation changes the flow features to a major extent. In the conventional direction of rotation the arm is aligned almost orthogonal to the upstream streamlines. In the reverse direction of rotation the blunt body is much more aligned with the flow streamlines. Figures 8.39 and 8.40 show instantaneous snapshots of the turbulent field in the drive for the conventional and reverse directions of rotation, respectively. One observes that in the conventional direction of rotation the expanding section (see Section 7.5.5) causes the flow to separate at the shroud and near the hub. This turbulent flow is then diverted back towards the hub by the arm and hence the suspension and slider are affected by large turbulent forcing. In the reverse direction the arm does not divert the flow towards the hub, but instead towards the base of the e-block arm. Turbulent forcing at this location is not as detrimental as at the suspension and arm – hence we expect the flow induced vibrations to be smaller in the reverse direction of rotation.

There are two additional observations: 1) The size of the wake is reduced in the case of reverse direction rotation. Vorticity shed from the arm quickly organizes into turbulent eddies and these eddies are transported towards the shroud by the spinning disks. 2) In the reverse direction of rotation there is considerable generation of turbulent structures upstream of the arm. These are partly generated at the tip of the high speed disk and partly at the upstream shroud expansion. These structures, however, do not affect the suspensions or sliders and are transported to the base of the e-block arm.

8.4.2   Drag on the actuator arm

A time history of the coefficients of drag on the arm (in the three directions: off-track, on-track and z) are shown in Figure 8.41, and the corresponding frequency spectra are shown Figure 8.42. These coefficients of drag are directly compared to the coefficients evaluated in Chapter 6, Figure 7.45. The sign of the $C_{D,off}$ and $C_{D,on}$ for the reverse direction is opposite to that for the conventional direction, hence the absolute values of the drag coefficients are plotted.

One immediately observes a great reduction in the fluctuation of the forcing with the reverse direction of disk rotation. The mean values are also shifted, higher in the off-track direction and lower in the on-track direction. The frequency spectra of Figure 8.42 show a major reduction in spectral content of the flow forcing. The spectra of the two simulation cases are very similar, except that the reverse direction of rotation is damped by approximately 5-7dB in the low frequency range, and the higher frequency base-line is damped by approximately 15dB.

A summary of the mean and RMS of the drag coefficients are shown in Figures 8.43 and 8.44 Our simulations show 58% reduction in the off-track RMS drag coefficient, 82% reduction in the on-track RMS drag coefficient and 58% reduction in the drag coefficient in the z direction.

Finally, we note that these percentages of reduction are generally higher than those reported in the experiments of Zeng and Hirano (2005). The reason for this is that our simulations can account for only one source of vibration – flow induced forcing. Other sources, such as spindle motor vibrations, disk runout or vibrations, rigid body motion of the actuator due to play at the actuator bearing, etc. (probably) do not vanish by reversing the direction of rotation and hence contribute equally to the vibrations of the arm. For this reason experimental studies must witness smaller reductions in actuator vibrations compared to our simulations.

8.4.3   Pressure forcing on the disk

One of the main motivations for computing the current simulation model (i.e. Case 2 from Chapter 7) versus the ones considered in the previous part of this Chapter was to be able to report the disk vibrations. The mean and RMS pressure distributions on the disks rotating in the conventional direction are shown in Figures 8.45 and 8.46, while the reverse direction are depicted in Figures 8.47 and 8.48. The mean pressure distributions are similar for both directions of rotation and are generally close to zero. The slight bias towards positive pressures around the disk is only an indicator of unconverged statistics of the flow. The RMS fluctuations of the pressure are similar in their magnitude, but the peaks in fluctuations occur at different locations. In the conventional direction almost all of the fluctuations are concentrated in the wake of the arm, while some larger peaks are evident near the shroud expansion. In the reverse direction of rotation large pressure fluctuations are observed at the edge of the disk, both upstream and downstream of the arm. The spectral content of the pressure fluctuations (discussed in detail in Section 7.5.3) is very similar in the two cases (largely because of the same grid and modeling conditions), and hence it is not discussed here.

Finally, the response of the disk to the pressure forcing is shown in Figure 8.49 for the conventional direction and Figure 8.51 for reverse spinning disks. This response has been calculated using the software code developed by in house and discussed in Chapter 6. In the mean, both cases show significant deformation to the (1,0) disk mode, which is also called the potato-chip mode. This mode is the lowest frequency mode and hence has the most energy stored in it.

The RMS vibrations are essentially axisymmetric and (close to) a linear function of the radial direction. More importantly the RMS vibrations are smaller in the reverse rotation case by approximately 35%. This is in good agreement with the data of Zeng and Hirano (2005).

8.5   Conclusions

This Chapter has been application oriented – we have shifted the focus from the simulation methods to their application to realistic disk drive flows.

The use of flow mitigation devices discussed above generally result in additional points of vorticity shedding, and depending on where the turbulence intensity is increased in the drive, this additional turbulence may or may not affect the actuator arm. On the other hand, M1 and M2 actually decrease the mean velocity of the flow as is demonstrated in the axial velocity profiles. This reduction in the kinetic energy of the flow (for the same disk rotation speed) causes reduced velocity fluctuations in the wake and in the regions immediately close to the actuator arm. Reiterating, in close proximity to the actuator arm, especially in the region of the base-plate and suspension (See points #11-23 in Figure 8.22) M1 has the smallest RMS in-plane and out-of-plane velocity fluctuations. M1 and M2 also have the smallest pressure fluctuations in this region while M3 appears to be a bad candidate based on all the RMS data presented.

We note that pressure-based loading on the actuator accounts for most of the off-track vibrations since pressure drag is 2 orders in magnitude larger than viscous drag. From this metric both M1 and M2 appear to be suitable candidates for reducing flow induced vibrations. However, we note that velocity fluctuations are also responsible for fluctuation of the forces on the actuator, and their effect may not appear directly in the RMS of the pressure fluctuations, which is a second order statistical moment. Changes in the velocity field near the arm cause changes in circulation around the arm, which is linearly related to the loading on the actuator arm. (For the Kutta-Zhukowski theorem, see Kundu (1990)) Taking this into consideration, it appears that M1 is a better candidate than M2 for reducing flow induced vibrations.

Finally, we also simulated the basic drive configuration with a reverse spinning disk. This apparently simple modification to the operation of the drive has the potential for very significant ramifications on the TMR. Our analysis shows that the coefficients of drag on the arm (only due to the flow) decrease by 50-80% and the vibrations of the disk reduce by approximately 35%. This simulation also provides computational confirmation of the experiments of Zeng and Hirano (2005).

The major portion of this work has now been presented in this and the preceding Chapters. The next and final Chapter will present the concluions.

8.6 Tables

Table 8.1: Geometry data

Number of disks	2
Number of e-block arms	1
Number of base plates	2
Number of suspensions	2
Number of sliders	2
Spacing between disks (mm)	2.2
Disk diameter (mm)	76.2
Width of shroud gap (mm)	1
Length of actuator (mm)	45
Length of e-block arm (mm)	32.5
Length of base plate (mm)	6.5
Length of suspension (mm)	11.1
Thickness of e-block arm (mm)	0.8
Thickness of base plate (mm)	0.3
Thickness of suspension (mm)	0.1
Dimensions of slider (mm)	$1 \times 0.8 \times 0.3$
Number of weight saving holes in e-block arm	2

]

Table 8.2: Model specific geometry data, M1

Thickness of blocking plate (mm)	0.8
Angular dimension of blocking plate (degress)	180
Radial dimension of blocking plate (mm)	16.25

Table 8.3: Model specific geometry data, M2

Thickness of downstream spoiler (mm)	1.6
Maximum width of downstream spolier (mm)	2.65
Length of downstream spoiler (mm)	20.75

Table 8.4: Model specific geometry data, M3

Thickness of upstream spoiler (mm)	1.4
Maximum width of upstream spolier (mm)	8
Length of upstream spoiler (mm)	17.5

Table 8.5: CFD modeling information

Governing equations	Filtered Navier Stokes equations
Solution algorithm	SIMPLEC Van doormaal and Raithby (1984)
Large eddy simulation model	Algebraic dynamic Germano et al. (1991)
Type of LES filter	Top-hat (variable width)
Temporal differencing scheme	Implicit Euler
Spatial differencing scheme (convective term)	Central differencing
Time step (seconds)	2.0×10^{-5}
Number of time steps	2400
Corresponding number of disk rotations	8
Initial conditions	Steady k-ϵ solution

Table 8.6: Boundary conditions

Disks	Rigid rotating walls, no slip
Shroud	Rigid wall, no slip
Shroud gap	Axial symmetry (zero normal gradient)
Other top and bottom surfaces	
of computational volume	Axial symmetry (zero normal gradient)
Hub/base of e-block arm	Fixed (similar to a cantilever)
Slider-disk interface	Slider slips on disk
	No cells between slider and disk
All structural interfaces	Rigidly joined
(e.g. suspension+slider,	(i.e. no dimple)
e-block arm+base plate)	
All fluid-structure surfaces	Coupled for pressure
	and shear stress

Table 8.7: Grid information

	M0	M1	M2	M3
Type of mesh	unstructured,	←	←	←
	quad-dominant			
Number of vols.	1,025,772	872,284	890,532	895,769
Max. cell vol. (mm^3)	8.996×10^{-2}	9.521×10^{-2}	9.436×10^{-2}	9.390×10^{-2}
Min. cell vol. (mm^3)	3.433×10^{-5}	3.315×10^{-5}	5.836×10^{-5}	6.855×10^{-5}
Avg. cell vol. (mm^3)	1.179×10^{-2}	1.243×10^{-2}	1.442×10^{-2}	1.571×10^{-2}
Avg. grid res. (mm)	0.2276	0.2316	0.2434	0.2504

Table 8.8: Common Legend for Figures in the paper

M0	Full Line	Original Simulation
M1	Dashed Line	Blocking Plate
M2	Dotted Line	Downstream Spoiler
M3	Dash-Dotted Line	Upstream Spoiler

8.7   Figures

Figure 8.1: Top view of M0: original simulation

Figure 8.2: Top view of M1: blocking plate

Figure 8.3: Top view of M2: downstream spoiler

Figure 8.4: Top view of M3: upstream spoiler

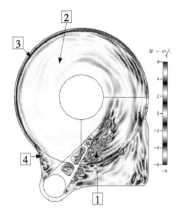

Figure 8.5: M0: Snapshot of turbulent field in the drive. Plot of axial velocity component on the midplane

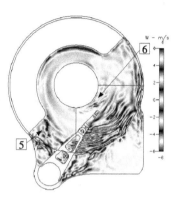

Figure 8.6: M1: Snapshot of turbulent field in the drive. Plot of axial velocity component on the midplane

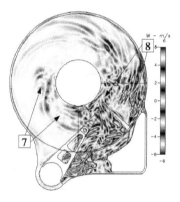

Figure 8.7: M2: Snapshot of turbulent field in the drive. Plot of axial velocity component on the midplane

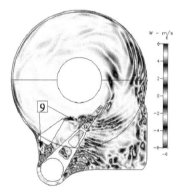

Figure 8.8: M3: Snapshot of turbulent field in the drive. Plot of axial velocity component on the midplane

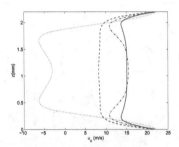

Figure 8.9: Inter-disk azimuthal velocity profile, at 340° from origin, i.e. in
the wake (See Table 8.8 for legend)

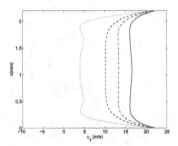

Figure 8.10: Inter-disk azimuthal velocity profile, at 45° from origin (See
Table 8.8 for legend)

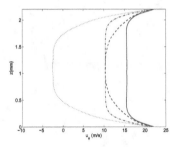

Figure 8.11: Inter-disk azimuthal velocity profile, at 135° from origin (See
Table 8.8 for legend)

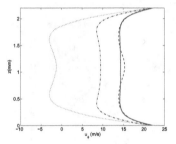

Figure 8.12: Inter-disk azimuthal velocity profile, at 225° from origin (See
Table 8.8 for legend)

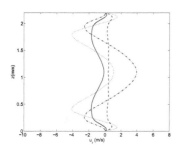

Figure 8.13: Inter-disk radial velocity profile, at 340° from origin, i.e. in the wake (See Table 8.8 for legend)

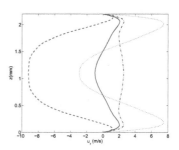

Figure 8.14: Inter-disk radial velocity profile, at 45° from origin (See Table 8.8 for legend)

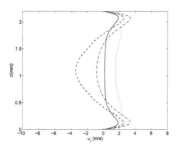

Figure 8.15: Inter-disk radial velocity profile, at 135° from origin (See Table 8.8 for legend)

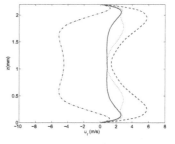

Figure 8.16: Inter-disk radial velocity profile, at 225° from origin (See Table 8.8 for legend)

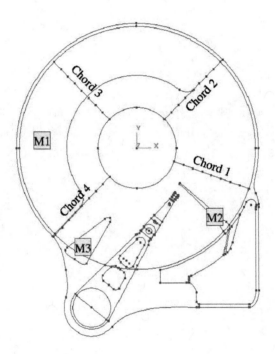

Figure 8.17: Chord locations for calculation of turblence intensity

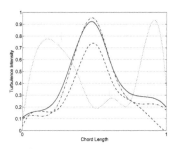

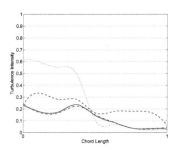

Figure 8.18: Turbulence Intensity along chord 1 (See Table 8.8 for legend)

Figure 8.19: Turbulence Intensity along chord 2 (See Table 8.8 for legend)

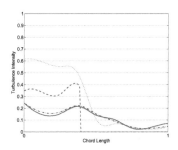

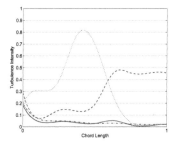

Figure 8.20: Turbulence Intensity along chord 3 (See Table 8.8 for legend)

Figure 8.21: Turbulence Intensity along chord 4 (See Table 8.8 for legend)

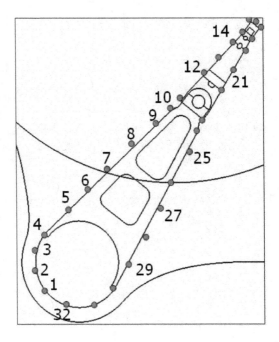

Figure 8.22: Location of points along actuator face for which velocity and pressure data is reported

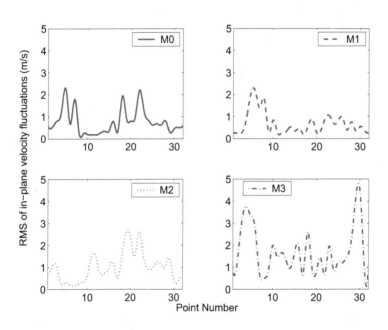

Figure 8.23: RMS fluctuation of in-plane velocity fluctuations (See Table 8.8 for legend)

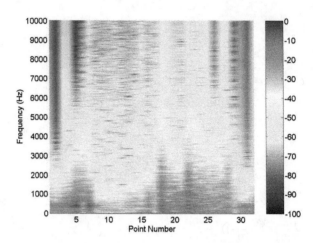

Figure 8.24: M0: Frequency Spectra of in-plane velocity fluctuations for data points 1-32

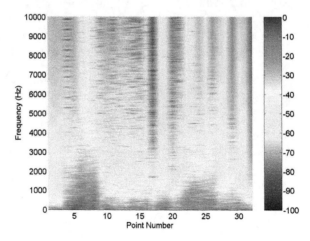

Figure 8.25: M1: Frequency Spectra of in-plane velocity fluctuations for data points 1-32

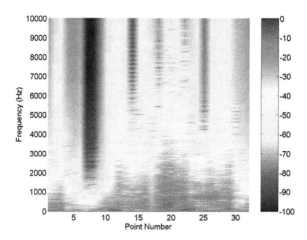

Figure 8.26: M2: Frequency Spectra of in-plane velocity fluctuations for data points 1-32

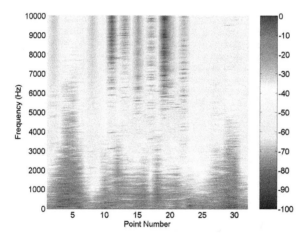

Figure 8.27: M3: Frequency Spectra of in-plane velocity fluctuations for data points 1-32

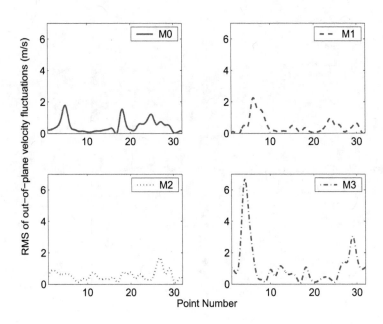

Figure 8.28: RMS fluctuation of out-of-plane velocity fluctuations (See Table 8.8 for legend)

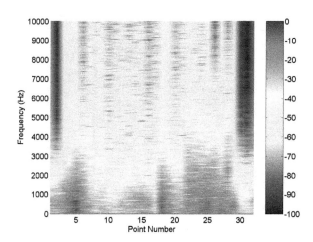

Figure 8.29: M0: Frequency Spectra of out-of-plane velocity fluctuations for data points 1-32

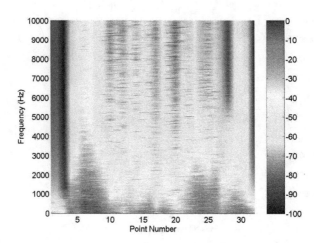

Figure 8.30: M1: Frequency Spectra of out-of-plane velocity fluctuations for data points 1-32

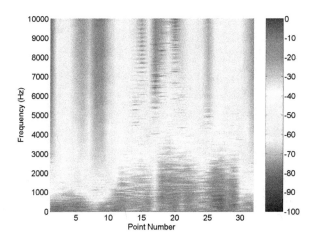

Figure 8.31: M2: Frequency Spectra of out-of-plane velocity fluctuations for data points 1-32

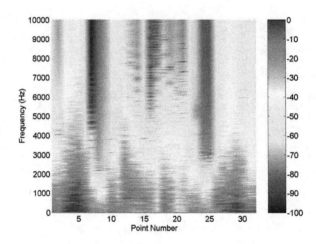

Figure 8.32: M3: Frequency Spectra of out-of-plane velocity fluctuations for data points 1-32

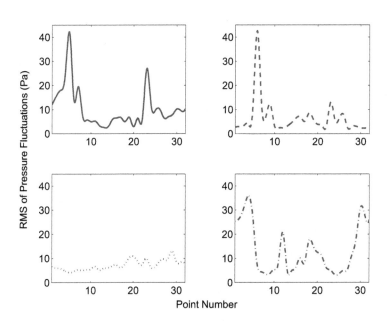

Figure 8.33: RMS fluctuations of Pressure (See Table 8.8 for legend)

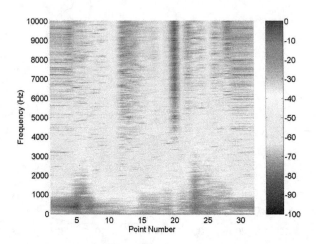

Figure 8.34: M0: Frequency Spectra of pressure fluctuations for data points 1-32

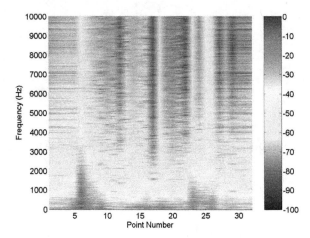

Figure 8.35: M1: Frequency Spectra of pressure fluctuations for data points 1-32

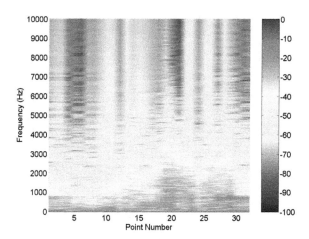

Figure 8.36: M2: Frequency Spectra of pressure fluctuations for data points 1-32

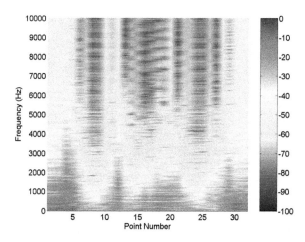

Figure 8.37: M3: Frequency Spectra of pressure fluctuations for data points 1-32

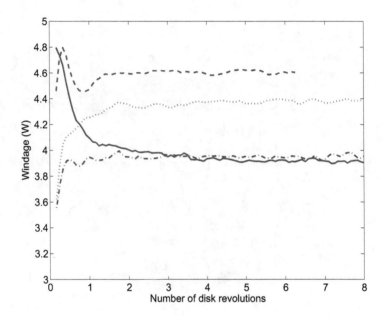

Figure 8.38: Windage loss at disks (See Table 8.8 for legend)

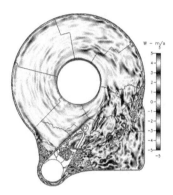 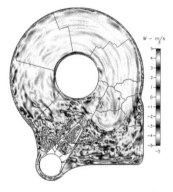

Figure 8.39: Snapshot of turbulent field in the drive with conventional direction of disk rotation. Plot of axial velocity component on a plane passing through the e-block arm.

Figure 8.40: Snapshot of turbulent field in the drive with reverse direction of disk rotation. Plot of axial velocity component on a plane passing through the e-block arm.

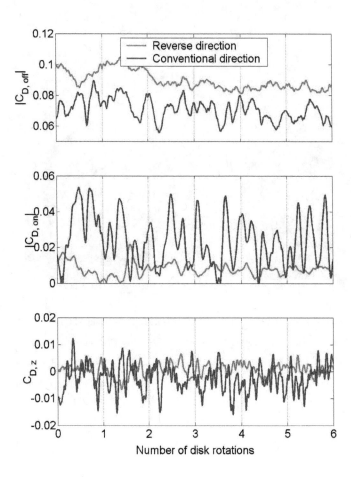

Figure 8.41: Time History of the coefficient of drag experienced by the arm, in three directions: off-track, on-track and z

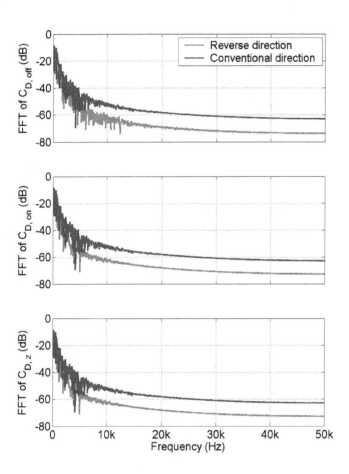

Figure 8.42: Frequency spectra of the coefficient of drag experienced by the arm, in three directions: off-track, on-track and z

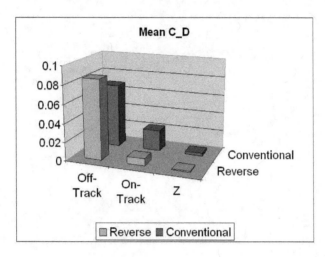

Figure 8.43: Summary of the mean coefficients of drag in the three directions: off-track, on-track and z

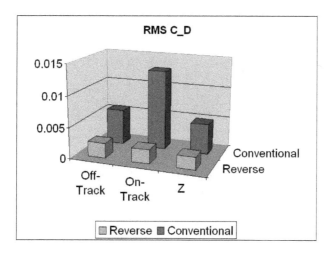

Figure 8.44: Summary of the rms coefficients of drag in the three directions: off-track, on-track and z

Figure 8.45: Resultant mean pressure in Pascals acting on the disks, when spun in the conventional direction.

Figure 8.46: Resultant rms pressure in Pascals acting on the disks, when spun in the conventional direction.

Figure 8.47: Resultant mean pressure in Pascals acting on the disks, when spun in the reverse direction.

Figure 8.48: Resultant rms pressure in Pascals acting on the disks, when spun in the reverse direction.

Figure 8.49: Mean disk deflection due to the airflow pressure, when the disk is spun in the conventional direction

Figure 8.50: RMS of the disk vibration due to the airflow pressure, when the disk is spun in the conventional direction

Figure 8.51: Mean disk deflection due to the airflow pressure, when the disk is spun in the reverse direction

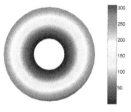

Figure 8.52: RMS of the disk vibration due to the airflow pressure, when the disk is spun in the reverse direction

Chapter 9

Conclusion

9.1 Conclusions

At the end of the previous Chapters we have described the conclusions of that Chapter in detail. In this final Chapter there appears no need to reiterate the conclusions to that level of detail. Hence only a general discussion of the topic follows.

It is probably accurate to say that the work presented in this book has advanced to some extent the state of the art of disk drive air flow simulations. While it has become relatively easy to compute such flows using commercial CFD software, demonstrating a sufficient level of confidence and sophistication in these simulations is not always straightforward.

As is evident the work presented in this book has been of three types:

- Solving the actual problem, i.e. computing the flow induced vibrations of the arm –
  which is an easy task of only marginal use unless it is defended by rigorous validation
  tests

- Testing, validating and developing confidence in our initial results, and

- Applying sophisticated simulation techniques beyond the flow induced vibrations of
  the arm, e.g. the vibrations of the disk, the use of reverse spinning disks, etc.

As we saw in Chapter 2 solving the actual problem of interest is relatively easy given the ease of use of commercial geometric modeling tools and commercial CFD solvers. It only requires a basic knowledge of CFD techniques and modern post processing tools are able to generate detailed plots of almost any desirable quantity. However, in solving the original problem of interest several questions need to be answered: How should the turbulence in the drive be modeled? If using LES, which SGS model is the best candidate? Among the commercial software available (and their in-built SGS models), which one provides the most accurate results? How do solutions change with the differencing scheme? What grid sizes should one use for reasonable results; and what is the uncertainty associated with those grids? What is the magnitude of the artificial dissipation? Chapters 3-6 have answered each of these questions in depth and will hopefully serve as a useful guide for future researchers in the field. Chapters 7-8 have focused on the third part of this book: applying the CFD techniques to some different problems, e.g. disk vibrations, flow mitigation devices and reverse spinning disks.

9.2 Future Work

Several research directions could serve as possible areas for the continuation and advancement of this work.

First and foremost, more realistic calculations should be performed to determine the response of the actuator arm to the flow. For such calculations it is imperative to model the air bearing separation at the slider using the compressible Reynolds equation and take into account the various non-linear forces that arise at small separations. Some of these are: van der Walls forces, meniscus forces, electrostatic forces and contact and impact forces when the slider (or a portion of the slider) touches the rotating disk. The model should also include the mechanical and electrostatic properties of the disk and other features such as the roughness of the disk. And finally, it is vital to include the elastic behavior of the suspensions, base plates and e-block arm.

As mentioned earlier, the development of such simulation capabilities are beyond the scope of this book. Nonetheless, such capabilities have been developed by others (notably contemporary researchers) and our CFD simulations would serve as a useful dataset for them. In the discussion that follows, we make use of such a code developed by our colleague: Vineet Gupta.

The next topic that deserves future attention is the cost of simulation. Throughout Chapters 3-8 we have mentioned the use of 32 - 128 CPUs in computing these flows to reasonable accuracy. In spite of this massive resource usage, simulation times for LES calculations are usually 2-3 weeks. Such large simulation times makes it infeasible (even impossible) to use CFD as a design tool and considerably limits the use of the CFD results.

As a possible solution to the above problem we propose a novel technique to quickly approximate certain CFD results without compromising their accuracy. This is presented in the next Section.

9.3 Use of Models to approximate forcing spectra

9.3.1 Introduction

As mentioned earlier the use of extensive computer resources to compute an LES result of a single flow field within 2-3 weeks may be impractical. In any case from previous Chapters we understand that most LES results are inherently accompanied by errors and uncertainties.

Modeling errors such as incompressibility, the SGS model and the boundary conditions are characteristic of all such computations and are hard to quantify. On the other hand, numerical errors such as grid discretization, time stepping errors and artificial dissipation are easier to quantify. We determined that such errors lead to 20-30% uncertainty in our results. We also realize that compensating for such errors may be infeasible given the high cost of the current calculations. The question then arises: Is it really worth computing LES calculations for different flow configurations when the uncertainty of the results is known and the errors cannot be practically reduced?

Evidently, the flow-induced drag is the most important information to determine the response of the actuator to the flow. This flow-induced forcing may be described either in terms of its statistics in the time domain or its spectrum in the frequency domain. Time domain signals are hard to characterize since one may only describe statistical moments like RMS or the fluctuating energy content. On the other hand, the frequency spectra provide rich information about the energy content at various frequency levels and determine the amount of energy stored in each vibration mode of the actuator arm. The zero-frequency DC component of the spectrum is unimportant – it accounts only for the mean displacement of the arm which can be easily compensated for by the servo control system.

For the reasons discussed above we propose to use a simple model to describe the drag spectrum. Such a model is completely manufactured based on CFD experience, but as we will see ahead, it provides a suitable approximation to the final results of a CFD calculation.

9.3.2 Piecewise linear model

Figure 9.1 shows a typical spectrum obtained for the off-track drag coefficient as the final result of an extensive CFD calculation. As a first approximation, one may think of fitting a simple piecewise linear model (PLM) to the spectrum as shown by the dotted line in Figure 9.2. In fitting the PLM to the CFD spectrum we have chosen the lines such that the area under the curves are equal. This implies that the RMS of the PLM-based and the CFD-based drag coefficients are equal – which further ensures that the flow-induced energy transfer to the actuator remains the same.

The question naturally arises as to how well the PLM approximates the CFD-based spectrum. Figure 9.3 shows a comparison between the off-track vibration results using the original CFD model and the PLM model (dashed line). These results have been calculated using the full air-bearing simulation code mentioned above. The results essentially show that the statistical characteristics of the PLM based forcing are very similar to the CFD-based forcing. The difference in the RMS of the vibrations is only 3%, while Figure 9.4 shows that in both cases the sway modes of the actuator are the dominant modes of the response. This result is encouraging: it implies that as long as the spectral energy is contant, one may modify the CFD spectra to a simple linear model and obtain almost the same results. The difference between the original CFD and PLM based results is so small (3%) that it appears unnecessary to investigate other model shapes (e.g. parabolic, exponential) that may approximate the original spectra better.

The next question that arises is of the sensitivity of the results to the forcing model. It is well known that the air bearing interface is highly non-linear and the dynamics of the suspension are weakly coupled in all directions. Hence we wish to investigate whether small changes in the spectra lead to large changes in the vibration results. Figure 9.5 shows five different PLM spectra (Case 1 to 5) that were investigated as forcing functions in the same air bearing code mentioned above. Case 1 has the same energy as the CFD spectra. In the remainder of the Cases, the location of the central point of the linear model is shifted upwards and downwards by 15 dB and to the left and right by 2 kHz. The different models considered here contain different amounts of energy and hence we expect differences in the response of the arm to these Cases.

Figure 9.6 compares the percentage change in RMS of the forcing function due to the model (in Cases 1-5) versus the percentage change in the RMS of the structural vibrations in the off-track direction. The chart basically shows that the relationship between the input-output is fairly linear – with a constant of proportionality between 0.8 and 1.4. This gives us confidence that under circumstances where the PLM spectra are not accurate in their total energy content (i.e. RMS), the resulting inaccuracies of the structural vibrations will not be disproportionately large, in spite of the non-linearities of the head-disk interface.

9.3.3   Parametric Investigation

The feasibility of the PLM approach has been demonstrated above. When using an energy conserving model, the errors in the final vibrations of the arm are very small – and the changes in the input spectra are linearly related to changes in the output RMS vibrations.

The flow-induced forcing spectra are highly dependant on the particular flow under consideration. The major parameters that influence the forcing spectra are:

- The Reynolds number of the flow, which is determined by:

    - The disk speed of rotation
    - The hard drive form factor
    - The disk to disk separation

- The geometry of the arm with respect to the flow,

    - The actual geometry of the arm (thickness, length, holes)
    - The radial position of the arm with respect to the flow

- Other factors, such as:

    - Geometry of the disk drive enclosure
    - The disk to shroud spacing
    - Spoiler devices, filters, desiccants

While it may be very expensive to quantify the effect of each of these parameters on the forcing spectra, we concentrate our efforts of only two parameters: disk RPM (which determines the Reynolds number) and radial position of the arm.

For this parametric study we conducted LES calculations of a typical disk drive model (used in Chapter 2) at three different RPMs (5400, 7200 and 10,000) and three different radial positions of the arm (ID, MD and OD). The locations of the arm in the different radial positions is shown in Figure 9.7. The relevant details of the simulations (grid, boundary and initial conditions) have been described in Chapter 2.

Before we present results of the parametric variation, we would like to outline the complexity of the task involved. For forces acting on the actuator, there are 6 degrees of freedom and correspondingly 6 forcing spectra (3 forces, 3 moments). Additionally the flow induced loading is a function of space and hence the frequency spectra of the loading vary along the length of the actuator. Finally, the spectra are heavily dependant on the parameters outlined above. Hence, it would take a gargantuan effort to investigate the spectra at all locations of the arm as a function of all parameters in each degree of freedom. In the following paragraphs we only investigate the trends as a function of two parameters: RPM and radial location, and it turns out that the trends are faithfully replicated over the entire length of the actuator.

Figures 9.8 and 9.9 show the variation of the spectra as a function of radial position and RPM, respectively. The figures show the variation in all degrees of freedom (i.e. forces and moments in the off-track, on-track and z direction). Remarkably, all the degrees of freedom show the same trends – i.e. the energy content of the spectra increase by going from ID to OD and by increasing the RPM. In Figure 9.8 one observes that the spectra for the forces show a clearer trend than those for the moments. Generally, the moments are 20-40 dB (i.e 1-2 orders of magnitude) smaller than the corresponding forces. The spectra in Figure 9.9 are truncated at different frequencies because a different time step was used in each simulation, leading to different sampling frequencies. In each simulation the ratio of the time step to the RPM was kept constant (2×10^{-5} seconds time step for 10,000 RPM). Interestingly, the same trends shown in Figures 9.8 and 9.9 are evident along the entire length of the actuator. The location of the forcing in Figures 9.8 and 9.9 have been chosen at random: Figure 9.8 shows the loads integrated over the suspension, while Figure 9.9 shows the loads integrated over the e-block arm.

The clear manifestation of such trends makes it possible to build PLM based spectra for each degree of freedom along all positions of the actuator. As an example, linear models are developed for the two parameters (RPM and radial location) for only a single force acting at a single location of the arm.

Figure 9.10 shows the variation of the frequency spectra for the force acting on the slider in the off-track direction, as a function of the radial position of the arm. (As mentioned above, we are considering a single direction and single location only). The PLM spectra based on energy conservation are also shown along with the CFD spectra. Finally, the three model spectra for the three radial positions are shown in Figure 9.11. The trend is very clear from the figure. The movement of the three points that define the piecewise linear model is also clearly visible and may be used for generating (postulating) spectra without actually performing the corresponding CFD calculation .

Similarly, Figure 9.12 shows the variation of the frequency spectra for the force acting on the slider in the off-track direction, as a function of the disk RPM. Again, PLM spectra based on energy conservation are fitted to these spectra and are plotted together in Figure 9.13.

9.3.4   Conclusions

From the Figures 9.11 and 9.13 it may now be possible to construct linear models to approximate conditions for which no CFD results are available. E.g. it is possible to approximate the forcing functions at different arm positions simply by linearly interpolating between the curves of Figure 9.11. On the other hand, different Reynolds number (either by form factor, disk RPM or disk-to-disk spacing) can be studied by interpolating (or extrapolating) the curves in Figure 9.13. All such simulations would totally bypass the CFD part of the solution strategy and hence would result in immense time and cost savings.

All such models are associated with some uncertainty. There could possibly be errors in the approximating PLMs that differ from the actual CFD results. However, given the clear trends demonstrated in our simulations, it is unlikely that the models will deviate from the CFD based results substantially. Moreover, as demonstrated earlier, slight differences in the energy content of the spectra lead to similarly slight differences in the final vibrations of the arm (i.e. errors do not blow up). As a final cautionary note: it is worth mentioning that the models developed here are based on the original CFD models. The errors and uncertainties of the underlying CFD data are obviously carried over to the PLM spectra.

9.4 Figures

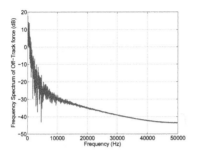

Figure 9.1: Typical flow induced forcing spectrum. Obtained for resultant force on the actuator acting in the off-track direction

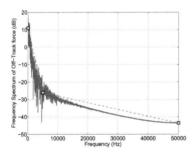

Figure 9.2: A piecewise linear model (PLM) of flow induced forcing spectrum shown along with the original CFD based spectrum

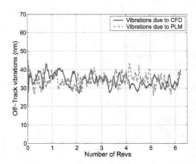

Figure 9.3: Comparison of the time history of the off-track vibrations of the slider when using the original CFD model v/s using the piecewise linear model (PLM)

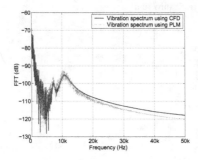

Figure 9.4: Frequency spectra of the slider off-track vibration showning the excitation of the first two sway modes at 6.69 kHz and 10.66 kHz

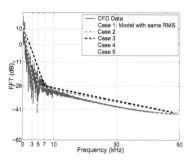

Figure 9.5: Investigation of the five different cases of PLM spectra, shown along with the original CFD spectrum

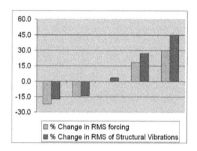

Figure 9.6: Chart showing the sensitivity of the off-track RMS results with respect to small changes in the forcing model spectrum

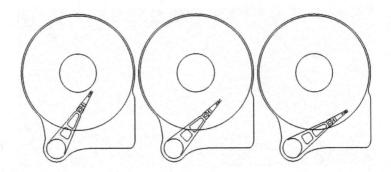

Figure 9.7: Computational models used for parametric investigation of radial arm position. Left: inner diameter (ID), center: middle diameter (MD) and right: outer diameter (OD)

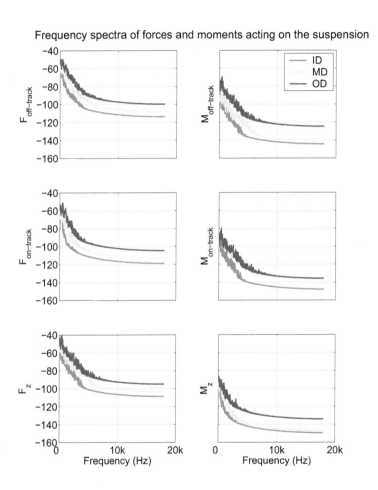

Figure 9.8: Variation of the spectra in all 6 degrees of freedom with radial position of the arm. Spectra are for the resultant forces and moments acting on the suspension

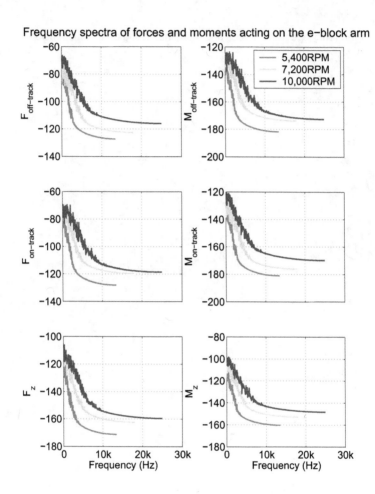

Figure 9.9: Variation of the spectra in all 6 degrees of freedom with disk RPM. Spectra are for the resultant forces and moments acting on the e-block arm

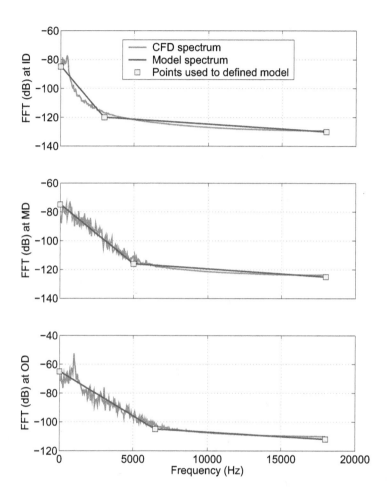

Figure 9.10: Variation of the frequency spectrum of the force on the suspension in the off-track direction as a function of the radial position. Also shown is the variation of the energy conserving PLM spectra

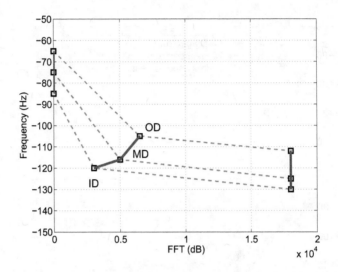

Figure 9.11: Variation of the PLM spectra due to change in radial position

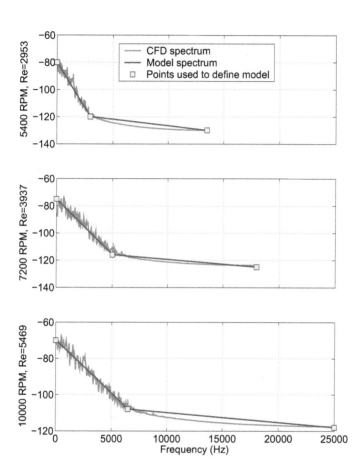

Figure 9.12: Variation of the frequency spectrum of the force on the suspension in the off-track direction as a function of the disk RPM. Also shown is the variation of the energy conserving PLM spectra

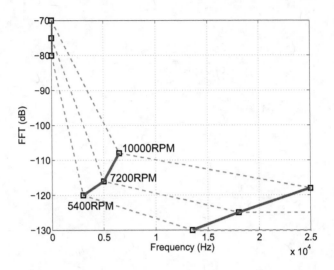

Figure 9.13: Variation of the PLM spectra due to change in disk RPM

Bibliography

S. D. Abrahamson, John Eaton, and D. J. Koga. The flow between shrouded co-rotating disks. *Physics of Fluids*, 1(2):241–251, 1989.

J. D. Jr. Anderson. *Computational Fluid Dynamics*. McGraw-Hill, 1995.

C. Barbier. *Experimental investigation into flows in model hard disk drives*. PhD thesis, University of Virginia, Charlottesville, 2006.

P. W. Bearman and E. D. Obasaju. An experimental study of pressure fluctuations on fixed and oscillating square-section cylinders. 119:297–321, 1982.

P. Bhargava and David B. Bogy. Numerical simulation of operational-shock in small form factor drives. *CML Report*, 05-005, 2005.

J. Boussinesq. Theorie de l'ecoulement tourbillant. *Mem. Pres. Acad. Sci. Paris*, 23:46, 1877.

C. J. Chang, J. A. C. Humphrey, and R. Grief. Calculation of turbulent convection between corotating disks in axisymmetric enclosures. *International Journal of Heat and Mass Transfer*, 33(12):2701–2720, 1990.

Y-B. Chang, D-K Park, N-C Park, and Y-P Park. Prediction of track misregistration due to disk flutter in hard disk drive. *IEEE Transactions on Magnetics*, (2):1441–1446, 2002.

J. Chung, J.-E. Oh, and H. H. Yoo. Non-Linear Vibration of a Flexible Spinning Disc with Angular Acceleration. *Journal of Sound Vibration*, 231:375–391, March 2000.

R. D. Cook, D. S. Malkus, and M. E. Plesha. *Concepts and Applications of Finite Element Analysis*. John Wiley and Sons, third edition, 1989.

C. D'Angelo. *Vibration and Aeroelastic stability of a Disk Rotating in a Fluid*. PhD thesis, University of California, Berkeley, 1991.

S. Deeyienyang and K. Ono. Suppression of resonance amplitude of disk vibrations by squeeze air be aring plate. *IEEE Transactions on Magnetics*, 37(2):820–825, 2001.

J. H. Ferziger. *Higher level simulations of turbulent flows in Compuational methods for turbulent, transonic and viscous flows*. 1983.

259

J. H. Ferziger. *Simulation and modeling of turbulent flows*. Cambridge University Press, 1996.

C. J. Freitas. Selected benchmarks from commercial cfd codes. *ASME Journal of Fluids Engineering*, 117:208–218, 1995.

R. Fukaya, S. Obi, S. Masuda, and M. Tokuyama. Flow instability and elastic vibration of shrouded corotating disk systems. *Experiments in Fluids*, 33:369–373, July 2002.

T. C. Fung. Numerical dissipation in time-step integration algorithms for structural dynamic analysis. *Progress in Structural Engineering and Materials*, (3):167–180, 2003.

C. Fureby, G. Tabor, H. G. Weller, and A. D. Gosman. Large eddy simulations of the flow around a square prism. *AIAA Journal*, 38(3):442–452, 2000.

C. Fureby, G. Tabor, H.G. Weller, and A.D. Gosman. A comparative study of subgrid scale models in homogenous isotropic turbulence. *Physics of Fluids*, 9:1416–1429, 1997.

M. Germano, U. Piomelli, P. Moin, and W. H. Cabot. A dynamic sub-grid scale eddy viscosity model. *Physics of Fluids*, A(3):1760–1765, 1991.

J. Girard, S. Abrahamson, and K. Uznanski. The effect of rotary arms on co-rotating disk flow. *ASME Journal of Fluids Engineering*, 117:259–272, 1995.

H. Gross. *Off-Track Vibrations of the Read-Write Heads in Hard Disk Drives*. PhD thesis, University of California, Berkeley, 2003.

L Guo and Y-J. D. Chen. Disk flutter and its impact on hdd servo performance. *IEEE Transactions on Magnetics*, 37(2):866–870, 2001.

F. Hendriks and A. Chan. The aerodynamic bypass in hard disk drives. *Proceedings of the ASME Information Storage and Processing Systems Co nference*, 2005.

Y. Hirono, T. Arisaka, N. Nishijima, T. Shimizu, S. Nakamura, and H. Masuda. Flow-induced vibration reduction in hdd by using a spoiler. *IEEE Transactions on Magnetics*, 40(4):3168–3170, 2004.

H. Hosaka and S. Crandall. Self-excited vibrations of a flexible disk rotating on an air film above a flat surface. *Acta Mech.*, 3:115–127, 1992.

J. A. C. Humphrey, C. A. Schuler, and D. R. Webster. Unsteady laminar flow between a pair of disks corotating in a fixed cylindrical enclosure. *Physics of Fluids*, 7(6):1225–1240, 1995.

I. Iglesias and J. A. C. Humphrey. Two- and three-dimensional laminar flows between disks co-rotating in a fixed cylindrical enclosure. *International Journal for numerical methods in fluids*, 26:581–603, 1998.

S. Imai. Fluid dynamics mechanism of disk flutter by measuring the pressure between disks. *IEEE transactions on magnetics*, 37(2):837–841, 2001.

R. Kaneko, S. Oguchi, and K. Hoshiya. Hydrodynamic characteristics in disk packs for magnetic storage. *Review of the Electrical Communication Laboratories, Nippon Telegraph and Telephone Public Corp., Japan*, 25:1325–1336, 1977.

N. Kang and A. Raman. Aeroelastic flutter mechanisms of a flexible disk rotating in an enclosed compressible fluid. *Journal of Applied Mechanics (Transactions of the ASME)*, 71(1):120–130, 2004.

H. Kazemi. *Mathematical Modeling of Flow-Induced Vibrations of Suspension-Head Units in Hard Disk Drives*. PhD thesis, University of Virginia, Charlottesville, 2004.

B. C. Kim, A. Raman, and C. D. Mote. Prediction of Aeroelastic Flutter in a Hard Disk Drive. *Journal of Sound Vibration*, 238:309–325, November 2000.

W. Kim and S. Menon. A new dynamic one-equation subgrid-scale model for large eddy simulation. In *33rd Aerospace Sciences Meeting and Exhibit*, Reno, NV, 1995.

P. J. Kundu. *Fluid Mechanics*. Academic Press, 1990.

D. Lakehal and F. Thiele. Sensitivity of turbulent shedding flows to non-linear stress-strain relations and reynolds stress models. *Computers and Fluids*, 30:1–35, 2001.

H. Lamb and R. V. Southwell. The Vibrations of a Spinning Disk. *Royal Society of London Proceedings Series A*, 99:272–280, July 1921.

B. E. Launder and B. I. Sharma. Application of the energy-dissipation model of turbulence to the calculation of flow near a spinning disc. *Letters Heat Mass Transfer*, 1:131–137, December 1974.

B. E. Lee. The effect of turbulence on the surface pressure field of a square prism. *Journal of Fluid Mechanics*, 69(2):263–292, 1975.

E. Lennemann. Aerodynamic aspects of disk files. *IBM J. Res. Develop*, pages 480–488, 1974.

D. K. Lilly. The representation of small-scale turbulence in numerical simulation experiments. In *BM Scientific Computing Symposium on Environmental Sciences*, Yorktown Heights, NY, 1967.

D. K. Lilly. A proposed modification of the germano subgrid scale closure method. *Physics of Fluids*, A(4):633–635, 1992.

D. A. Lyn, S. Einav, W. Rodi, and J.-H. Park. A laser-doppler velocimetry study of ensemble-averaged characteristics of the turbulent near wake of a square cylinder. *Journal of Fluid Mechanics*, 304:285–319, 1995.

D. A. Lyn and W. Rodi. The flapping shear layer formed by flow separation from the forward corner of a square cylinder. *Journal of Fluid Mechanics*, 267:353–376, 1994.

I. McLean and I. Gartshore. Spanwise correlations of pressure on a rigid square section cylinder. 41-44:779–808, 1992.

S. Menon and W. Kim. Application of the localized dynamic subgrid scale model to turbulent wall-bounded flows. *AIAA Paper*, 97-0210, 1997.

R. Mittal and P. Moin. Suitability of upwind-biased finite difference schemes for large-eddy simulation of turbulent flows. *AIAA Journal*, 38:1415–1417, 1997.

S. Nakamura, S. Wakatsuki, T. Haruhide, S. Saegusa, and Y. Hirono. Flow-induced vibration of head gimbal assembly. *IEEE Transactions on Magnetics*, 40(4):3198–3200, 2004.

E. Y. K. Ng and Z. Y. Liu. Prediction of unobstructed flow for co-rotating multi disk drive in an enclosure. *International Journal for numerical methods in fluids*, 35:519–531, 2001.

F. Nicoud and F. Ducros. Subgrid-scale modelling based on the square of the velocity gradient tensor. *Flow, Turbulence and Combustion*, 62:183–200, 1999.

C. Norberg. Flow around rectangular cylinders: Pressure forces and wake frequencies. *Journal of Wind Engineering and Industrial Aerodynamics*, 49:187–196, 1993.

U. Piomelli. High reynolds number calculations using the dynamic subgrid scale stress model. *Physics of Fluids*, A(6):1484–1490, 1993.

U. Piomelli, P. Moin, and J. H. Ferziger. Model consistency in the large eddy simulation of turbulent channel flows. *Physics of Fluids*, (31):1884–1891, 1988.

S. B. Pope. *Turbulent Flows*. Cambridge University Press, 2003.

W. Rodi, J. H. Ferziger, M. Breuer, and M. Pourquie. Status of large eddy simulation: Results of a workshop. *ASME Journal of Fluids Engineering*, 119:248–261, 1997.

R. S. Rogallo and P. Moin. Numerical simulation of turbulent flows. *Annual Review of fluid mechanics*, 16:99–137, 1984.

C. A. Schuler, W. Usry, B. Weber, J. A. C. Humphrey, and R. Grief. On the flow in the unobstructed space between shrouded corotating disks. *Physics of Fluids A 2*, 10: 1760–1770, 1990.

H. Shimizu, T. Shimizu, M. Tokuyama, H. Masuda, and S. Nakamura. Numerical simulation of positioning error caused by air-flow-induced vibration of head gimbals assembly in hard disk drive. *IEEE Transcations on Magnetics*, 39(2):806–811, 2003.

H. Shimizu, M. Tokuyama, S. Imai, S. Nakamura, and K. Sakai. Study of aerodynamic characteristics in hard disk drives by numerical simulation. *IEEE Transcations on Magnetics*, 37(2):831–836, 2001.

J. Smagorinsky. General circulation experiments with the primitive equations, i. the basic experiment. *Monthly Weather Review*, 91:99–164, 1963.

R. Smirnov, S. Shi, and I. Celik. Random flow generation technique for large eddy simulations and particle-dynamics modeling. *ASME Journal of Fluids Engineering*, 123: 359–371, 2001.

A. Sohankar, L. Davidson, and C. Norberg. Large eddy simulation of flow past a square cylinder: Comparison of different subgrid scale models. *ASME Journal of Fluids Engineering*, 122:39–47, 2000.

H Song, M. Damodaran, and Q. Y. Ng. Simulation of flow field and particle trajectories in hard disk drive enclosures. *High Performance Computation for Engineered Systems (HPCES)*, Jan 2004.

F. Stern, R. V. Wilson, H. W. Coleman, and E. G. Paterson. Comprehensive approach to verification and validation of cfd simulations – part 1: Methodology and procedures. *ASME Journal of Fluids Engineering*, 123:793–802, 2001.

H. Suzuki and J. A. C. Humphrey. Flow past large obstructions between corotating disks in fixed cylindrical enclosures. *ASME Journal of Fluids Engineering*, 119:499–505, 1997.

M. Tatewaki, N. Tsuda, and T. Maruyama. A numerical simulation of unsteady airflow in hdds. *FUJITSU Sci. Tech. J.*, 37(2):227–235, 2001.

J. W. Thomas. *Numerical Partial Differential Equations: Finite Difference Methods.* Springer, 1998. ISBN 0387979999.

N. Tsuda, H. Kobutera, M. Tatewaki, S. Noda, M. Hashiguchi, and T. Maruyama. Unsteady analysis and experimental verification of the aerodynamic vibration mechanism of hdd arms. *IEEE Transcations on Magnetics*, 39(2):819–825, 2003.

H. M. Tzeng and J. A. C. Humphrey. Corotating disk flow in an axisymmetric enclosure with and without a bluff body. *International Journal of Heat and Fluid Flow*, 12:194–201, 1991.

W. R. Usry, J. A. C. Humphrey, and R. Grief. Unsteady flow in the unobstructed space between disks corotating in a cylidrical enclosure. *ASME Journal of Fluids Engineering*, 115:620–626, 1993.

J. P. Van doormaal and G. D. Raithby. Enhancements of the simple method incompressible fluid flows. *Numerical Heat Transfer*, 7:147–163, 1984.

E. R. Van Driest. On turbulent flow near a wall. *Journal of Aero. Science*, 23:1007–1011, 1956.

B. J. Vickery. Fluctuating lift and drag on a long cylinder of square cross section in a smooth and in a turbulent stream. *Journal of Fluid Mechanics*, 25:481–494, 1966.

P. R. Voke. *Direct and Large Eddy Simulation II*, pages 355–422. Kulwer Acadamic Publishers, 1997.

B. Vreman, B. Geurts, and H. Kuerten. Large-eddy simulation of the turbulent mixing layer. *Journal of Fluid Mechanics*, 339:357–390, 1997.

Wikipedia.   Early   ibm   disk   storage,   2006.   URL `http://en.wikipedia.org/wiki/Early_IBM_disk_storage`.   [Online; accessed 27-March-2006].

R. Wood, J Miles, and T. Olson. Recording technologies for terabit per square inch systems. *IEEE Transactions on Magnetics*, 38(4):1711–1718, 2002.

Y. Yamaguchi, A. A. Talukder, T. Shibuya, and M. Tokuyama. Air flow around a magnetic head slider suspension and its effects on the slider flying height fluctuations. *IEEE Transactions on Magnetics*, 26(5):2430–2432, 1990.

Z. X. Yuan, A. C. J. Luo, and X. Yan. Airflow pressure and shear forces on a rotating, deformed disk in an open shroud. *Communications in Nonlinear Science and Numerical Simulation*, 9(5):481–497, 2004.

Q. Zeng and T. Hirano. Some experimental observations of disk drives with reverse spinning disks. In *ASME Information Storage and Processing Systems Conference*, June 2005.

Wissenschaftlicher Buchverlag bietet

kostenfreie

Publikation

von

wissenschaftlichen Arbeiten

Diplomarbeiten, Magisterarbeiten, Master und Bachelor Theses
sowie Dissertationen, Habilitationen und wissenschaftliche Monographien

Sie verfügen über eine wissenschaftliche Abschlußarbeit zu aktuellen oder zeitlosen
Fragestellungen, die hohen inhaltlichen und formalen Ansprüchen genügt,
und haben **Interesse an einer honorarvergüteten Publikation**?

Dann senden Sie bitte erste Informationen über Ihre Arbeit per Email
an info@vdm-verlag.de. Unser Außenlektorat meldet sich umgehend bei Ihnen.

VDM Verlag Dr. Müller Aktiengesellschaft & Co. KG
Dudweiler Landstraße 125a
D - 66123 Saarbrücken

www.vdm-verlag.de

www.ingramcontent.com/pod-product-compliance
Lightning Source LLC
LaVergne TN
LVHW022304060326
832902LV00020B/3261